Decision-making in High-Risk Projects

Akio Torii

Decision-making in High-Risk Projects

Springer

Akio Torii
Yokohama National University
Yokohama, Kanagawa, Japan

Institute of Economic Research
Chuo University
Hachioji, Tokyo, Japan

ISBN 978-981-97-3055-1 ISBN 978-981-97-3056-8 (eBook)
https://doi.org/10.1007/978-981-97-3056-8

This Springer imprint is published by the registered company Springer Nature Singapore Pte Ltd.
The registered company address is: 152 Beach Road, #21-01/04 Gateway East, Singapore 189721, Singapore

Preface

This book addresses the complex decision-making issues societies face when considering projects that promise significant benefits, such as nuclear power generation, but also pose risks of severe accidents. It explores the societal dilemma of whether to advance or halt such initiatives. The book is divided into two parts.

The first part discusses the decision-making process for projects like nuclear energy, which can cause severe damage in the event of an accident. Risk assessment must be conducted well in advance. However, the nuclear disaster at TEPCO (Tokyo Electric Power Company)'s Fukushima Daiichi Power Station on March 11, 2011 raises questions about the adequacy of risk-assessment actions. This part investigates the potential factors that lead to insufficient risk assessment in the decision-making process. A model is developed wherein the business entity responsible for operating the project invests in obtaining risk-assessment information, while the decision-making entity makes decisions based on its perceived balance of social costs and benefits. The analysis identifies several factors causing insufficient risk assessment: optimism about the risks involved on the side of the operating business entity, large private benefits associated with the project, a gap between the magnitude of the risks the operating entity perceives and the standards of judgment required by society, and ambiguous liability for compensation. Factors that do not deter the operating entity from effort in risk assessment include the ability to practically make its own decisions by capturing the decision-making entity and the ability to keep risk-related information confidential.

The second part aims to identify the difficulties in decision-making in such situations by creating and analyzing a decision-making model for organizations that advance high-risk projects through trial and error. Even projects with significant risks may need to be considered for implementation given the expected benefits to society. In such examinations, it is crucial to determine whether it is possible to avoid a severe accident (i.e., to assess the extent of the risk). In most cases, it is common to estimate technical risks while trialing the project and make decisions incrementally. However, executing this step-by-step decision-making is challenging. The objective of this part is to analyze the difficulty of incremental decision-making. Recognizing the characteristics of such decision-making challenges is crucial to avoid severe accidents and prevent erroneous decisions.

This study initially embarked on a different trajectory, focusing on empirical industrial organization. My interest was piqued by the peculiar characteristics of industries related to nuclear power generation, such as their multiple subcontracting system. Despite being generally perceived as technologically advanced, these supporting industries appeared inefficient. I was intrigued by the factors impeding their efficiency, which I believed could influence the energy industry's operation. However, the Fukushima Daiichi Nuclear Power Plant disaster prompted a reevaluation of my research direction, leading to a temporary halt. As time passed without a clear research direction, debates on the structural issues behind Japan's nuclear industry following the accident gradually subsided, and nuclear power operations resumed sequentially, with plants reopening one by one. It remains questionable whether policy measures adequately reflect the lessons learned from the major accident and whether the energy structure dependent on nuclear power is being reconsidered.

Motivated by this context, this study was initiated to understand why no efforts have been made to reach a public consensus on the merits and demerits of nuclear power reliance, despite extensive research and debate. The study employs mathematical model analysis, with models constructed to depict the situation leading to the accident at TEPCO's Fukushima Daiichi power station. However, the study does not aim to use the model to identify the institutional factors in the Japanese energy industry that contributed to the accident. Instead, it seeks to answer the following question: Why is nuclear power being promoted without sufficient discussion and research? While I believe further research on this issue is necessary, I present my current analysis here. Numerous research questions remain.

I express my gratitude to Chuo University for providing the opportunity to conduct this research. I also thank the participants in the Marketing Theory, Industrial Organization, and Business Economics Workshop held at Nanzan University's Center for Management Studies, and Professor Teizou Anayama, who discussed the analysis developed in this book at the international conference by the Japan Economic Policy Association. I am grateful to Springer Nature publishing for the opportunity to publish research results, as well as to the editor, Mr. Yutaka Hirachi. Lastly, I thank Editage for their diligent proofreading of the authors' challenging English text.

A part of this work was supported by JSPS KAKENHI Grant Number 18K01582.

Hachioji, Japan Akio Torii

Contents

Abbreviations

DM	Decision-Maker
HRO	High-Reliability Organization
JAERI	Japan Atomic Energy Research Institute
JPSR	JAERI's Passive Safety Reactor
NAIIC	Fukushima Nuclear Accident Independent Investigation and Verification Commission
PIUS	Process Inherent Ultimate Safety reactor
TEPCO	Tokyo Electric Power Company

Mathematical Notations

$\varnothing$	Empty set
B	Benefit of the project to all members of society
$(\,)^c$	Complementary set
$c(\cdot)$	Cost of effort
$\bar{c}$	Upper bound on $c(\cdot)$
D	Damage caused by an accident
$\bar{D}$	Critical expected accidental damage beyond which the expected project value for society members increases with an observation
D^M	DM's estimate of accident damage
D^A	Agent's estimate of accident damage
D_1^A, D_2^A	Boundaries separating the regions of D^A where the yield from the agent's second trial is positive
D_+^M, D_+^A	Critical expected accidental damage for the DM and the agent, respectively, beyond which the agent's project value increases with the second observation
$\bar{D}^A$	Critical expected accidental damage of the agent beyond which the agent's expected project value increases with the first observation

e	Agent's level of effort in investigation trial
$\bar{e}$	Upper limit of e beyond which no increase of the probability π
e^*	Effort level chosen by the agent
e_i	Agent's effort level chosen for Case i in Chap. 6
$\mathcal{F}$	σ-algebra of all measurable events within the sample space Ω
$F(\cdot)$	Cumulative distribution function of parameter θ
$f(\cdot)$	Probability density function of parameter θ
$H_\sigma(\cdot)$	Indicator function
n	Number of observations
P	Probability measure on $(\omega, \mathcal{F})$
$p(\cdot)$	Probability function
R	Private profit accrued by the agent
$\mathbb{R}$	The set of all real numbers
$\mathbb{R}^+$	The set of all non-negative real numbers
$\mathbb{R}^{++}$	The set of all positive real numbers
$U(\cdot)$	Uniform distribution
V_i^M, V_i^A	Expected project value for the DM and the agent for Case i in Chap. 6
$W(\cdot)$	Project value function
W_0^A	Project value for the agent, based on subjective probability of accident occurrence with a no-information prior distribution
W_0^M	Project value for the DM, based on subjective probability of accident occurrence with a no-information prior distribution
$W_1^a(\cdot)$	Project value for the agent, contingent upon the observation of a precursor
W_1^A	Expected project value for the agent, based on the subjective probability of observing one precursor
W_n	Expected project value for society members when n observation is given
X	Random variable that governs the occurrence of an accident
x_1^A	Optimal critical value of the first observation for the agent beyond which the project should be halted
x_i	Observed random variable representing the outcomes of the i-th trial
$\boldsymbol{x}$	Collective term referring to $\boldsymbol{x}_0$, $\boldsymbol{x}_1$, and $\boldsymbol{x}_2$
$\boldsymbol{x}_0$	Zero-dimensional vector representing the absence of any observation ($\boldsymbol{x}_0 = \varnothing$)
$\boldsymbol{x}_1, \boldsymbol{x}_2$	Vectors representing the specific realizations of the outcomes ω, for one or two observations, respectively
x_n^*	Optimal critical value of the observation for the DM beyond which the project should be halted when n observation(s) are given
$Y(\cdot)$	Payoff for society members
Y_1^a	Agent's expected project value when the second investigation trial is allowed
Y_2^a	Agent's payoff during the second trial
y_n	Highest value of n observation(s)
β_n	Probability of the type II error when n observation(s) is(are) given
γ_n	Probability of the type I error when n observation(s) is(are) given

θ	Parameter of uniform distribution
λ	Constant probability when the effort is no less than $\bar{e}$
$\pi(\cdot)$	Probability a precursor of accident is provided
ρ	Decision where project execution is effectively decided in each case when the observed precursor can be kept concealed
Σ	Collective term referring to Σ^0, Σ^1, and Σ^2
Σ^i	Specialized σ-algebra for scenarios with i observation(s)
σ	Subset of Σ representing the condition under which the DM decides to execute the project
σ^*	Optimal σ for the DM
σ^A	Optimal σ for the agent
σ_i	Subset of Σ^i representing the condition under which the DM decides to execute the project
σ_1^A	Decision that optimizes the project value for the agent point-wisely
τ	Threshold value beyond which an accident occurs
τ^+	Critical threshold of τ beyond which the agent's project value increases with the second observation
$\Phi(\cdot)$	Function used to define the expected project value in Chap. 6
χ_1, χ_2	Boundaries separating the regions of x_1 where the condition the agent exerts effort for the second observation
Ω	Sample space of all possible outcomes of the observations
ω	Individual outcome within the sample space

Part I
Risk-Assessment Behavior and Decision-Making Mechanisms in Projects with the Potential for Severe Accidents

Chapter 1
Introduction: Decision-Making Problems in Risky Projects

1.1 Fukushima Nuclear Accident in 2011

This part analyzes the research activity of an agent who is delegated to proceed with a high-risk project with the potential for a severe accident. Recognizing that, in reality, projects with a high risk of severe accidents proceed without sufficient research activities on the risks,[1] I investigate here what factors lead to such insufficient risk assessment in the decision-making process.

Some hold the view that the nuclear disaster at the Fukushima Daiichi Nuclear Power Plant on March 11, 2011, following the Great East Japan Earthquake could have been prevented had appropriate measures been implemented. For example, Saito (2015, Chap. 7) scrutinizes the interim and final reports of the Fukushima Nuclear Accident Independent Investigation and Verification Commission (NAIIC) and concludes that "the accident was within the scope and extent of what had been anticipated, and measures to prevent the accident from spreading were also agreed upon in advance. To borrow a legal expression, the accident was both foreseeable and avoidable." Many others have expressed the view that the accident could have been avoided.[2] Furthermore, the interim report of the NAIIC states that at the Tokai No. 2 Power Station of the Japan Atomic Power Company, which faced a similar possibility of a severe accident, the power supply necessary to cool the reactor was secured,

---

[1] The content of this chapter and 2 has been expanded from earlier publications, Torii (2017), Torii (2022).

[2] For example, Fuchigami and Iino (2016) conclude that at least 2–3 years ago, and as early as a decade ago, material existed that predicted a huge tsunami would follow a huge earthquake; if a giant tsunami had been predicted in advance, countermeasures could have been implemented relatively easily within the timeframe.

A. Torii, *Decision-making in High-Risk Projects*,
https://doi.org/10.1007/978-981-97-3056-8_1

because the measures were based on the "Ibaraki Prefecture Tsunami Inundation Zone Map" prepared by the Ibaraki Prefecture in 2007.[3] Indeed, it is natural to reason that the accident could have been avoided or the damage caused by the accident could have been substantially lessened if the high risk had been accurately assessed, countermeasures for possible risks had been considered, and appropriate measures had been adopted in accordance with the procedures prepared in advance based on an accurate assessment of the situation.

Why, then, was the risk not properly assessed? The basic attitude of the government toward nuclear power before the accident is shown in "the Report of the Subcommittee on Nuclear Energy of the Electricity Industry Committee of the Advisory Committee for Natural Resources and Energy—'Nuclear Power Nation Plan'." While the report proposes the strategic promotion of nuclear power generation and the nuclear power industry, it makes little mention of the risks associated with the promotion of nuclear power generation. The report simply mentions a brief discussion of operational and other safety regulations as a "reference" at the end of the report. There is no indication of any attempt to assess risk and formulate policy to deal with the risk.

The National Diet of Japan Fukushima Nuclear Accident Independent Investigation Commission concluded that the root cause of the accident was the "collapse of the monitoring and supervisory function of nuclear safety due to the 'inversion' of the relationship between the regulators and Tokyo Electric Power Company (TEPCO)." Indeed, the capture theory, by which a regulated entity can take advantage of information superiority or the existence of private information to capture regulators, has been widely presented since Stigler et al. (1971).

However, even if the hypothesis that the regulators were captive to the electric utilities were true, does it mean that the electric utilities perceived the risks associated with their business as low, were they less risk-averse, or both? This would have led them to attempt to avoid the measures necessary to promote nuclear energy business as costly or leading to the loss of profit opportunities. Regulators, meanwhile, would have either recognized the risks as high or been more risk-averse, and thus, would have been willing to take measures against the possibility of accidents, even if they were expensive. In this case, the utility companies would have successfully lobbied the regulators and succeeded in making them captive and getting them to endorse their behavior of ignoring the risk and avoiding countermeasures. Alternatively, were the electric utilities not interested in the risks associated with their projects, instead holding optimistic expectations without any particular basis for them? In this case, the high cost of assessing the risks associated with the project would have caused them to avoid the act of assessing the risks themselves. They would have captured

[3] According to the Investigation Committee on the Accident at the Fukushima Nuclear Power Stations (2011), "The wave height of the tsunami caused by the Great East Japan Earthquake at Tokai No. 2 Power Station was estimated to be T.P.+5.4 m. Since the sidewall penetration work was not completed, seawater flooded into the pump room from the penetration area, causing the emergency. Although one emergency diesel generator unit was shut down, the remaining two generators were able to provide the power necessary to cool the reactor because the sidewall height had been increased from T.P.+4.91m to T.P.+6.11m" (p. 407), where T.P. is the mean sea level of Tokyo Bay.

the regulators that were trying to assess the risks and got them to tacitly approve their attitudes to avoid risk assessment. The assertion that private operators are more likely to ignore risk than regulators is merely a hypothesis and has not been empirically verified.

In general, risk assessment is an important management decision when promoting a project that requires a large investment. In Japan, it is often argued that managers are generally risk-averse and, therefore, underinvest.[4] Ito (2013) argues that if the government had not taken responsibility, private power companies would not have entered the dangerous business of nuclear power. Ito further insists that TEPCO was operating under the strong regulation of the Agency for Natural Resources and Energy and, therefore, went ahead with its business development without regard to risk. It is hypothesized that TEPCO, as a private company, should have evaluated risks before making decisions, but instead, TEPCO assumed that it was insulated from risks by regulations, which led it to pursue extremely risky businesses despite its inherently risk-averse tendencies. Ito further explains the relationship between the electric power companies and the regulatory agency: "In accordance with the Japanese practice of the government not taking responsibility, the electric power companies apply for approval, which is then granted. In effect, however, the Agency for Natural Resources and Energy creates the basic policy, and the power companies simply apply for approval based on that policy" (Ito 2013, p. 102).[5]

Another view is that the opponents and proponents of nuclear power have confronted each other without stepping back, which has increased the risks associated with nuclear power generation itself. "How to proceed with nuclear power reform" (Special Task Force for Nuclear Power Reform of TEPCO 2012) lists the second factor behind the lack of severe accident countermeasures as the concern that acknowledging the need for severe accident countermeasures would lead to litigation risks, and the third factor regarding the adoption of severe accident countermeasures would arouse anxiety in the local community and the public, which would give nuclear opponents momentum. The existence of the opposition movement prevented TEPCO from adopting severe accident countermeasures that would have been implemented if there had been no opposition movement. There are many views of this sort. Takeda (2011), for example, argues that the risk has increased as a result of nuclear power

---

[4] John et al. (2008) and Acharya et al. (2011) measure the level of corporate risk-taking by the standard deviation of earnings before interest, taxes, depreciation, and amortization per total assets, and they show that Japanese firms' risk-taking is remarkably low. Arikawa et al. (2016) introduce studies that examine the tendency of Japanese firms to be pessimistic, which hinders aggressive investment. Furthermore, the Cabinet Office (2008) states that risk-taking by Japanese companies is not sufficient and is a challenge for increasing growth potential. This indicates that the Japanese government views its own companies as risk-averse.

[5] The view that bureaucrats have wide discretion by means of administrative guidance to achieve de facto coercion is considered the "bureaucratic regulation theory," which has been analyzed by Yamauchi (1974), Institute of Administrative Management (1981, 1982a, b), and others. See also Tatebayashi (1994). Yamauchi (1974) holds that informal restraining measures effectively threaten to force agencies to submit to regulatory guidance by implying abuse of authority.

continuing to be generated in the face of the strong opposition movement.[6] Whether the existence of the nuclear opposition movement actually had the effect of increasing the risk is a hypothesis that needs to be verified. However, it is true that uneasy apprehension exists in society. This study is not concerned with the question of whether a specific opposition movement had the effect of increasing risk. However, I consider it worth thoroughly examining the effect of the opposition movement's demand for a strict risk tolerance level imposed from outside the operating agency on actual risk-assessment behavior.

1.2 Overview of the Model Analysis

This part presents a model constructed to analyze the problem of inadequate risk-assessing behavior. The model describes the mechanisms and context of decision-making in nuclear power generation in Japan. The players in the model are the operating agency, which is in a position to assess risk because it is in charge of the project, and the regulator who has formal decision-making authority. In the model, I analyze the proposition that the manner of effective decision-making shared by both parties practically could have caused them to inhibit risk-assessing behavior. For example, the hypothesis is examined that the agent captures the regulator, which would have hampered adequate risk assessment due to the de facto shift of decision-making authority from the regulator to the operator. I examine another hypothesis that the agent acts without regard to the risks because it is operating in an environment in which the opposition is so strong that very strict risk tolerance is set in making decisions to proceed with the project, and even minor signs of an accident would shut down the project, as long as there were signs of an accident. Furthermore, the model analyzes how the private profits in operations by the agent, the liability, and

[6] Takeda (2011) states that, as a result of the continued use of nuclear power amid strong opposition, "when the proponents were unable to acquire land for new nuclear power plants due to obstruction from the opposition, they ended up building more plants on existing sites and crowded them together. This rather increased the risk of the Fukushima nuclear power plant" (p.31). He also states that "The Japan Atomic Energy Research Institute (JAERI) started research in 1993 to explore reactor types, such as the JPSR, that could be used as an extension of the PIUS reactor while ensuring passive safety. However, the results were hardly reported. Why? If we admit that there is a safer reactor, we would be admitting that the nuclear power plants currently in operation are inadequate. The existence of 'safer reactors' is unacceptable, and there is no way to choose one, as they have repeatedly stated that nuclear power plants in operation are 'absolutely safe' to reassure the public who are feeling uneasy." (JPSR is JAERI's Passive Safety Reactor while PIUS reactor refers to the Process Inherent Ultimate Safety reactor). Whether the strength of such opposition actually contributed to increased risk requires verification. For example, when additional power generation facilities are built at a single site, the construction cost for the first unit is often the highest when differences in construction timing and generating capacity are considered; thus, cost savings may be a factor for the high concentration of nuclear power generation. Research on new reactor types is also being promoted in Japan, and some third-generation reactors with enhanced safety features are already in operation (Yamaji 2009).

the possibility of information confidentiality affect the behavior and level of effort to assess risk.

The basic setup of the model is as follows. It is assumed that the entity that operates the project that may cause a severe accident is separate from the entity that makes decisions on whether to promote the project. This setup refers to the electric power company and the Agency for Natural Resources and Energy in the case of Japan. The two entities are called the agent and Decision-Maker (DM). The agent can search in advance, at a cost, for the level of risk associated with the project. If the agent puts effort into this search, it can obtain information about the likelihood of an accident occurring with a probability corresponding to its level of effort. However, information is not always provided. Furthermore, the level of effort cannot be directly observed and controlled. In this model, the determination of the level of effort is analyzed. The model separately examines the case in which information about the risk of an accident occurring can be observed by every member of society and the case in which only the agent can observe information, and selectively disclose it. When information is observable to every member of society, they use that information to update their own expectations about the risk they have. To update this expectation, updating using Bayes' theorem is assumed. Since everyone can use the same information, it is assumed that the forecast updates are shared by all members of society. Meanwhile, with regard to the expected damage value in the event of a severe accident, no objective distribution of damage value is assumed, but each member of society is assumed to have their own estimate of the amount of the damage. This estimate is the belief of each member, and there are optimistic and pessimistic members.

1.3 Main Conclusions

Explaining the main conclusions in this part should enhance reader comprehension.

The fundamental principle guiding this conclusion is that the entity executing a project should rationally determine the level of effort to be invested in risk assessment. This decision should be based on a comparison between the potential enhancement in business value resulting from risk assessment and the cost of the assessment. The increase in business value is achieved by reducing the variance in the probability of a severe accident, which subsequently decreases the variance in net return—the difference between the expected benefits and the potential damages in the event of an accident. As the ability to forecast risks more accurately improves, so does the precision in predicting net returns. Making decisions regarding business execution based on these more accurate forecasts—particularly by halting operations when signs of a negative net return emerge—improves the expected net return, thereby increasing business value. However, because the decision-making entity that determines whether the business proceeds differs from the entity that conducts the business and expends effort on risk assessment, the expected business value diverges in terms of benefits and damages (i.e., the extent of private benefits for the executing entity and the difference in their respective beliefs about the amount of damage). Hence, their criteria for desirable judgment differ. As the business entity undertakes the risk-assessment effort, the incentive for this effort is influenced by how the value of the business to the entity changes as a result of acquiring new information for risk assessment under the given system. The extent to which this additional information alters the business value for the entity represents the value of the information itself. That is, the value of additional information determines the incentive for effort in risk assessment.

The main conclusions of the model analysis are as follows:

First, the higher the extent of accident damage expected by the business entity and the smaller the private benefits associated with business operations, the higher the level of effort will be. This suggests that insufficient risk-assessment efforts could be attributed to overly optimistic entities or those pursuing private gains. Naturally, a high estimate of accident damage strengthens the incentive to assess risks and avoid potential accidents. Assuming decisions are made based on prior expectations in the absence of risk information, leading to the execution of projects, then implementing risk assessment could potentially halt the project. Therefore, in situations where private benefits are substantial, the increased risk of project cessation given risk assessment could somewhat impair the incentive to conduct prior risk assessments.

Second, assuming a constant expected damage amount for the business entity, the level of effort peaks when the DM is slightly more optimistic than the business entity, contingent on the size of the private benefits. If the DM sets a stringent standard, the business entity becomes less proactive in risk assessment. In other words, overly strict external safety standards result in a passive approach to risk assessment. When the judgment criteria of the DM and the business entity diverge beyond a level corresponding to the presence of private benefits, the value of risk-related information decreases. This decrease, in turn, leads to a reduction in the incentive to exert effort in risk assessment. This suggests that demanding stricter standards, such as through opposition movements, paradoxically leads to neglect in risk assessment. It is suggested that captivating the regulatory agency and imposing the business entity's risk judgment standards always increases risk-assessment effort in the absence of private benefits. Even with private benefits, captivating the regulatory agency increases risk-assessment efforts, provided the regulatory agency is more pessimistic than the business operators. A decrease in effort due to captivation only occurs when the regulatory agency is slightly more optimistic than the business operators, and this optimism is proportional to the size of the private benefits.

Third, when a business entity makes its own decisions, the level of effort increases with the business operator's pessimism about the amount of accident damage. This conclusion is derived from the characteristics discussed in the first and second points. When business operators make their own decisions about promoting their business, entrusting the business to more risk-averse companies can lead to higher efforts in risk assessment.

Fourth, imposing strict liability on a business entity increases the level of effort. This discussion does not concern negligence liability, which pertains to insufficient risk-assessment effort, but a strict liability, which addresses accident occurrence. Given that compensation liability functions in a manner that is equivalent to negative private benefits in decision-making, it appears as though the business entity's expected accident damage value increases by the compensation liability amount. This suggested that exempting business operators from liability could lead to a neglect of effort. If business entities believe they will be exempt from liability in the event of an accident, this belief could discourage active engagement in risk assessment.

On the other hand, fifth, when DM are subject to strict liability, the business entity's level of effort might increase if the decision-making entity is significantly more optimistic than the business entity and the compensation amount is small. However, if the decision-making entity is more pessimistic than the business entity, the business entity's level of effort will decrease. This is derived from the effect equivalent to the case where the decision-making entity's expected accident damage value increases by the compensation liability amount. Therefore, when the decision-making entity is more optimistic than the business entity, imposing moderate compensation liability on the regulatory side may lead to a more proactive risk assessment by the business entity. Conversely, when the decision-making entity is more pessimistic, demanding compensation from it reduces the business entity's effort. If it is clear that the Agency for Natural Resources and Energy, which makes the decisions, takes full responsibility for accident, and if this agency is more pessimistic than the electric power company, this can lead to neglect in risk assessment.

Sixth, when it is possible to conceal information, the ability to do so results in the business entity's level of effort either increasing or remaining the same, provided the business is perceived to bring societal benefits that exceed the expected value of accident damage. The level of effort might decrease in the specific scenario where the DM halts the business given a lack of information, though this is not always the case. In situations where information is unavailable, the decision-making entity cannot determine whether the information is truly unobtainable or merely concealed. This uncertainty can lead to two outcomes: the business may either halt or continue. If the business halts due to a lack of information, the business entity might underestimate or overestimate the potential accident damage compared to the decision-making entity. If the business entity underestimates the damage, it will increase its effort to obtain information to prevent business cessation. If it overestimates the damage, concealment might align the decision-making entity's decision with that of the business entity, thereby increasing the business value and the value of information, leading to increased effort. In scenarios where the business continues despite the absence of information, the business entity might again underestimate or overestimate the potential accident damage compared to the decision-making entity. If it underestimates the damage, concealment can align the decision-making entity's decision with that of the business entity, increasing both the business value and the value of information, thereby increasing its effort. If it overestimates the damage, the business will proceed even with concealment, and the decision-making entity's decision remains unchanged in a situation where concealment is not possible. Therefore, the ability to conceal information does not necessarily lead to a lax attitude toward risk. The only situation where a business entity might neglect risk-assessment efforts due to the ability to conceal information is when the information is not obtained, regulators suspect concealment and stop the project, and the business entity deems the business too risky. In this case, the business entity may intentionally neglect efforts to halt the business. Apart from this scenario, the concealment of information does not lead to a passive approach to risk assessment. In other words, the potential for concealment, identified as a communication problem between the business entity and the decision-making entity, does not decrease the effort in risk assessment.

The remainder of this part is organized as follows. Chapter 2 describes the relationship of the developed model to the existing literature. Chapter 3 describes the basic structure of the model. Chapter 4 reports the results of the analyses and obtains the propositions for the case in which every member of the society can observe the information, that is, when there is no private information. Chapter 5 investigates the case in which the agent can conceal what it finds in the search. Chapter 6 concludes and discusses the results in Part 1.

References

Acharya, Viral V, Yakov Amihud, and Lubomir Litov 2011."Creditor rights and corporate risk-taking." *J. Financ. Intermed.*. 102 (1): 150–166.

Arikawa, Yasuhiro, Kotaro Inoue, Takuji Saito, and Yohei Nagao 2016."Factors behind the low performance of Japanese firms: An international comparison." In *Corporate Governance and Growth Strategy* edited by Hideaki Miyajima: Toyo Keizai Shinpo Sha, in Japanese, Chap. 12 397–427.

Fuchigami, Masao, and Kenji Iino 2016."Study Group on Tsunami Countermeasures at Fukushima Nuclear Power Plant, Final Report." in Japanese.

Institute of Administrative Management 1981."Research Report Study on Administrative Guidance (1)." *Public Administration Review Quarterly*. (15): 36–54, in Japanese.

Institute of Administrative Management 1982a."Research Report Study on Administrative Guidance (2)." *Public Administration Review Quarterly*. (18): 38–51, in Japanese.

Institute of Administrative Management 1982b."Research Report Study on Administrative Guidance (3)." *Public Administration Review Quarterly*. (20): 37–48, in Japanese.

Investigation Committee on the Accident at the Fukushima Nuclear Power Stations 2011."Interium Report." http://www.cas.go.jp/jp/seisaku/icanps/post-1.html, in Japanese.

Ito, Mitsuharu 2013. *Political Economy of Nuclear Generation.* : Iwanami Shoten, in Japanese, in Japanese.

John, Kose, Lubomir Litov, and Bernard Yeung 2008."Corporate governance and risk-taking." *The journal of finance*. 63 (4): 1679–1728.

Saito, Makoto 2015. *Political Economy of Earthquake Reconstruction –Separation and Intersection of Tsunami Disaster and Nuclear Crisis–.* : Nihon Hyoron Sha, in Japanese.

Special Task Force for Nuclear Power Reform of TEPCO 2012."How to proceed with nuclear power reform." in Japanese.

Stigler, George J et al. 1971."The Theory of Economic Regulation." *Bell Journal of Economics*. 2 (1): 3–21.

Takeda, Toru 2011. *Nuclear Coverage and Media.* : Kodansha, in Japanese.

Tatebayashi, Masahiko 1994."Industrial Policy and Administration." In *Policy and Administration* edited by Masaru Nishio, and Muramatsu Michio 3 of Seminar Administration: Yuhikaku, in Japanese, Chap. 3.

the Cabinet Office 2008."Annual Report on the Japanese Economy and Public Finance 2008 – Japan's Economy: Tackling Risks —." in Japanese.

Torii, Akio 2017."Risk Assessing Behavior and Decision-making Mechanism in Risky Project Where Severe Accident May Occur." Institute of Economic Research Chuo Universiry Discussion Paper No.282, in Japanese.

Torii, Akio 2022."Severe Accident Avoidance and Risk Assessment—Institutional Factors that Hinder Risk Assessment Activities." In *Public Regulation Law and Policy* edited by Makoto Kurita, and Masashi Takeo 49 of Boissonade Institute of Modern Law and Politics Monograph Series: Hosei University Press, in Japanese, Chap. 4.

Yamaji, Kenji 2009. *The Past, Present, and Future of Nuclear Energy: Will Nuclear Power Be Restored?.* : Corona Pubulishing, in Japanese.

Yamauchi, Kazuo 1974."Administrative Guidance and the Rule of Law." *Jurist.* (506): 14–19, in Japanese.

Chapter 2
Literature Review

This part analyzes information-gathering behaviors for decision-making regarding the promotion of nuclear energy, as introduced in Chap. 1. Specifically, it focuses on the existence of decision-making entities and entities expected to gather information for conducting business and assessing risks in the process. Using a model, it aims to analyze how the institutional relationship between the two parties influences the behavior of risk assessment. A body of research on communication between informed experts and uninformed DMs has investigated the optimal decisions for DMs while relying on the agent's information. In particular, studies that analyze the incentives for experts to spend costs to acquire information address issues similar to those addressed in this study. As such, the model constructed in the prior literature is similar to the model in this study. However, the model in this study differs in several respects for the purpose of decision-making based on risk assessment. The model analysis in this study is not a simple application to decision-making problems involving nuclear energy. On the contrary, it is a new theoretical contribution in that some aspects of the model analysis have been developed in ways that have not been addressed by conventional model analysis. This chapter briefly introduces models from the existing literature and positions the model of this paper within those model analyses. Accordingly, the unique aspects of our model relative to others will become clear.

2.1 Endogenous Information Acquisition and Cheap Talk

The communication game between an informed expert and an uninformed DM is the subject of a series of cheap-talk models, starting with Crawford and Sobel (1982). The cheap-talk model is used to analyze communication to overcome inefficiencies that arise when there is information asymmetry.[1] While cheap-talk models usually assume that experts have exogenous expertise, Dur and Swank (2005) assume that

[1] For a comprehensive survey of the cheap-talk model, see, for example Sobel (2013).

A. Torii, *Decision-making in High-Risk Projects*,
https://doi.org/10.1007/978-981-97-3056-8_2

experts receive signals probabilistically depending on their effort level to analyze the relationship between incentives to acquire information and the content of communication. The signal is noisy, but the higher the effort, the lower the noise. Upon receiving this signal, the expert advises the DM on the pros and cons of the project. Cheap-talk models assume that experts and DMs have different preferences about project execution. Dur and Swank first show that experts who work hardest are neutral experts who have no preferences about project execution. Furthermore, they analyze the desirable experts from the DM's perspective. If the DM is biased, using an unbiased, neutral expert would yield advice based on information with less noise. However, given differing preferences, the advice might deviate from the DM's interests. Nevertheless, employing an expert biased similarly to themselves ensures the advice aligns with the DM's interests. However, owing to the expert's lower effort, the advice is based on information with more noise. They analyze this trade-off and conclude that it is optimal for the DM to use an expert who is somewhat biased in the same direction as themselves. Furthermore, Dur and Swank analyze cases where information is disclosed as hard evidence. In such cases, the issue of advice deviating from the DM's interests does not arise. The only concern becomes the expert's level of effort. Indeed, experts biased in a direction opposite to the DM's bias exert the most effort. As the DM is biased, experts anticipate that, by acquiring information, they can correct the DM's views. Experts biased in the opposite direction to the DM expect larger benefits from acquiring information favorable to their stance, thereby increasing their incentive to exert effort. Therefore, experts with preferences biased in a direction different from the DM are more motivated to diligently acquire information.

In Van den Steen (2010), the information acquisition effort decision is similarly analyzed. The agent's effort gives one sample of the project's probability of success. Using this sample, the agent and DM update the prior distribution of success probabilities. It is shown that the more the parameters of the prior distributions of the agent and the DM differ, the more the agent's incentive to obtain a new sample tends to increase. This effect is called the persuasion effect. It represents the tendency for the incentive to persuade others to be stronger the more their beliefs differ—hence the term "persuasion effect." This mechanism works because the more different the parameters of the prior distribution are, the greater the effect of the success or failure samples in updating the prior distribution. The more the agent perceives the DM to have extreme parameters, the wider the range of parameters the DM is expected to update with the new sample. This effect results in a greater increase in expected profits obtained from the information provided by the sample.

However, the section of Van den Steen (2010) analysis focusing specifically on persuasion effects does not address the issue of communication, and does not explicitly analyze the trade-off between the effects of differing views on communication and information gathering. Che and Kartik (2009) analyze the problem of information gathering using the cheap-talk model, which explicitly considers communication, and show that the factor they call the "avoiding prejudice effect" must be considered in addition to the persuasion effect. In their model, the agent receives a signal indicating the state of the world with a probability that depends on the agent's effort.

By presenting a signal, it exhibits a persuasion effect that encourages the updating of the prior distribution regarding the state of the world. The more biased the agent's prior distribution is the greater the incentive to acquire information owing to the persuasion effect. Here, it is assumed that the agent can keep the signals it obtains confidential, including their existence. This information concealment causes communication loss. To cope with this problem, the DM chooses an extreme behavior to punish the agent if the information is not reported. To avoid this, agents increase their efforts to obtain information. This is the avoiding prejudice effect. Che and Kartik (2009) further demonstrate that when there is a bias in preferences, this bias and differences in the prior distribution parameters interact synergistically. That is, when the differences in prior distribution and preferences are in the same direction, the persuasion effect is strengthened. Conversely, when they are in opposite directions, the effect is weakened.

The reason that the biased expert works harder in Dur and Swank (2005) is similar to, but different from, the persuasion effect. In Van den Steen (2010) and Che and Kartik (2009), a biased agent is defined as an entity that holds a biased prior distribution regarding the state of the environment. The persuasion effect is generated by a technical mechanism, where the addition of information to the DM's prior distribution leads to a more significant update of the distribution, especially when the new information greatly differs from the existing prior distribution. On the other hand, biased experts in Dur and Swank (2005) are those who have higher gains when decisions are made that differ from the DM's prior preferences, and they have greater incentives to acquire information that possibly reverses the DM's decisions.

In the model analysis in Chap. 4, all information acquired is assumed to be observable to all members of society. Furthermore, with respect to the occurrence of risk, the prior distribution of all entities is assumed to be identical to a non-informative prior distribution. However, the amount of damage in the event of an accident is considered given for each individual as their prior beliefs. For this reason, the persuasion effect does not arise. Nevertheless, the results show that agents with larger expected accident damage values exert more effort. As in Dur and Swank (2005), this is a manifestation of the effect that the benefit is much greater when the decision is reversed. Dur and Swank (2005) also derive the proposition that non-biased agents work harder. This conclusion depends on the assumption that information cannot be verified and does not hold if the information is verifiable and publicly available. This effect is from the maximization of the value of information to change decisions. This study shows that this proposition holds even under public information, as the less biased the agent is from the DM's point of view, the more effort the agent puts in. This effect occurs because the value of information to change the decision is maximized. This is exactly the same effect claimed by Dur and Swank (2005).

Most communication games between the informed expert and the uninformed DM follow the tradition of the cheap-talk model, in which knowledge is represented by a one-dimensional scalar number and the benefits produced by the knowledge are symmetric in deviations from the true value in both directions. Thus, whether the true value is overestimated or underestimated, symmetric losses or disutility occur. Therefore, the effect of bias on the expert is also symmetric. Meanwhile, the state of the environment in this study pertains to the risk of accident occurrence. The risk of an accident has two dimensions: the probability of an accident occurring and the amount of damage when an accident occurs. In the model analysis in Chap. 4, it is assumed that information on the probability of an accident occurring is shared, that everyone has the same prior distribution, and that the prior distribution is updated in the same way by the information obtained. Meanwhile, each member is assumed to hold a unique belief regarding the expected amount of damages in the event of an accident, implying that these beliefs vary per member. A once-off severe accident is considered here and a true value of accident damage is not assumed to exist. Then, the prediction of accident damage amounts is optimistic or pessimistic, and then asymmetric. Bias is also not taken as an absolute concept. Relative differences in beliefs between members are considered to be biased. Under these assumptions, we can confirm both the effect that biased agents work harder under public information, as in Dur and Swank (2005), and the effect that non-biased agents work harder under cheap-talk communication. The latter effect, in particular, is shown to be independent of the setting of cheap-talk communication. The expectation of accident damages is one-dimensional and monotonic. Nevertheless, similar results are obtained in that symmetric effects emerge with respect to whether the DM's expectation is larger or smaller than the agent's expectation. Chapter 6 of this part addresses scenarios where information can be concealed. Such scenarios have four distinct cases, each leading to either a pooling equilibrium or a separating equilibrium. The analysis demonstrates the ability to examine various states. This diversity largely stems from the model's capability to represent more varied situations given the risk information being two dimensional.

Another study on information acquisition behavior under cheap-talk game is Argenziano, Severinov and Squintani (2016), who analyzes cases in which information is obtained more than once and shows that DMs can use expert information to take more accurate actions even when the expert is biased. This is because DMs can punish experts by engaging in behavior called bubbling, which is ignoring information. Given this property, even if the division of the task of information gathering and decision-making into two separate entities causes communication loss, it can still be rational to deliberately separate these functions. This study also analyzes the pros and cons of separating and integrating information gathering and decision-making. Integration here means that decision-making is delegated to the executing entity. The behavior of the DM punishing the agent to encourage effort is not analyzed. However, the effect of the agent's capturing the DM is analyzed as the effect of the integration of the two entities. The level of effort is determined by the relative size relationship between the expected accident loss values of the two entities.

Gerardi and Yariv (2008) analyze the case of using multiple experts, which not only encourages the acquisition of information but also allows the information to be used as effectively as possible for the DM's decision-making. They analyze the trade-off between increasing the incentive to acquire information by employing experts close to the DM's preferences and increasing the disadvantage of making a wrong decision by employing experts far from the DM's preferences. Given a choice, the two experts who are most different from the DM in their view of which direction the error is more serious are selected. It is shown that the sequential use of these two experts leads to the most desirable decision, and that designing sequential decision-making, rather than simply using the two experts together, maximizes effort and allows for more accurate decision-making. Chapter 5 examines sequential information acquisition, emphasizing the continuous data collection by a single entity. The results suggest that the core characteristics align closely with those identified in isolated research activities. This chapter also investigates the inherent difficulties associated with sequential surveying processes.

2.2 Advocacy

The cheap-talk model highlights communication loss between the DM and the expert as a significant concern. Consequently, the information shared might not directly stem from the expert's findings but from advice based on such information (Dur and Swank 2005); it might even be concealed (Che and Kartik 2009). This study presupposes that the information is inherently verifiable and accessible to everyone, though specific issues regarding nuclear energy might necessitate the withholding of risk information. While cheap-talk-style analyses often consider verifiable public information as a model variant, this aspect does not constitute the primary focus of their analysis.

Research investigating the function of advocates analyzes models in which a DM with no expertise delegates the acquisition of information to an expert with expertise and primarily considers issues other than communication when the information is verifiable. Dewatripont and Tirole (1999) analyze the problems for a DM that delegates information acquisition for the purpose of making a decision to move from the status quo to one of two opposing states. Rewards to agents are assumed to align with the outcomes of a verifiable decision by the DM, not with unverifiable information acquisition behavior. Thus, if the DM delegates the task of information gathering to a single agent, and this agent discovers information favoring one side of a decision, followed by additional information favoring the opposing side, the DM opts to maintain the status quo. This sequence of obtaining conflicting information about the opposing sides related to the DM's potential decision leads to a decrease in rewards, thereby diminishing the agent's incentive. The resulting decrease in incentives can be compensated by increasing rewards when advantageous information is obtained by either party, which generates rent. The problem of rent is resolved by delegating information acquisition to two separate agents in different directions. In this sense, delegating information gathering to two conflicting agents is efficient. This is known as the system of advocacy.

Because the analysis illustrates the efficiency of delegating information acquisition to different agents for two opposing options, it has many implications for the nuclear energy debate, in which the acquisition of information involved in decision-making for opposing options is important. However, efficiency in this analysis is efficiency only in the sense that the reward to the agent can be minimized. In the debate on nuclear energy, the efficiency of the cost of information gathering is not as important as the benefits of information, since huge damages are expected to occur when the risk becomes a reality. The issue is how to provide incentives for agents to make efforts to obtain information.

In Dewatripont and Tirole (1999), the probability of information acquisition does not vary continuously with the level of effort, as in the cheap-talk model of information acquisition behavior, and the cost of effort is fixed. Therefore, the analysis seeks a remuneration structure that satisfies the incentive compatibility for information gathering for each of the two options. Thus, they analyze the lowest cost remuneration scheme that satisfies the incentive rationale. In this study, we are not interested in the design of the reward system but construct the model on the assumption that the probability of successful information acquisition varies continuously with the agent's level of effort. In Dewatripont and Tirole (1999), further analysis is conducted on various cases, including when the probability of information acquisition is considered an indicator of the agent's capability, when career concerns come into play, information can be concealed, and instances of self-advocacy.

Beniers, Dur and Swank (2002) analyze a case in which the information-gathering process of two agents is sequential in two phases: For the second phase of the investigation to be carried out, it must also reward the acquisition of information in a direction that negates the previously obtained information. Even if the status quo is maintained, a reward must be paid. However, to ensure the incentive for conducting the initial survey is not lost, additional rent must be paid. To deal with this problem, Kim (2014b) analyzes the case in which agents are rewarded for the information itself. Kim (2014b) argues that when the probability of collecting information is sufficiently low, compensation can be saved using agents with private interests in decisions in either direction. Kim (2014b) explains that using a single agent increases the rent of information collection, but using two agents avoids this problem.

Kim (2014a) defines the case in which one agent is delegated to gather information in two directions as an inquisitorial system by a judge, and the case in which two representatives are delegated in opposite directions as an adversarial system between a prosecutor and a defendant, and evaluates each based on sentencing error. Kim (2014a) concludes that the probability of making an error in judgment is lower in the adversarial system. When a judge gathers information, the incentive is tied to the extent to which the evidence contributes to their judgment. By contrast, when a prosecutor or a defendant collects information, the focus is on amassing evidence to win the case. The judge's incentive is proportional to the incremental probability brought about by new information, while the prosecutor's or defendant's incentive is proportional to the probability level of incorporating new information. This distinction can elicit greater effort for a prosecutor and a defendant, allowing fewer errors in the adversarial system.

In particular, Kim (2014a)'s criterion of how much effort can be elicited to minimize the likelihood of error has much to offer in considering nuclear energy issues. This is because the biggest challenge is to avoid errors in the decision on whether nuclear power should be promoted or halted. However, this study differs significantly from the conditions in Kim's model in that access to information is asymmetric.

Assessing the risks associated with nuclear energy requires not only a high level of expertise but also knowledge gained through actual operations. Thus, professional competence is asymmetrically distributed. It would be extremely difficult for opponents of nuclear energy, who anticipate the high dangers of nuclear energy, to gather information in accordance with their claims, compared to the information gathered by proponents. Even if the same degree of information were to occur in both directions, it is unlikely that information contrary to the promotion of nuclear energy would be disclosed. This problem is also analyzed in Dewatripont and Tirole (1999) as a problem of confidentiality of information in the advocacy system.

Furthermore, in the case of this type of model, it must be considered that the DM is set up as an entity that seeks to achieve social justice by weighing the information against the two options. The purpose of this study is to discuss how the relationship between the regulator, which in effect makes the decisions, and the operating company, which operates under regulation, affects the incentives to conduct risk assessments. For this purpose, the model's construction of a neutral DM and two opposing directions might not be appropriate, even if the DM is assumed to be biased in the preference. For this reason, this study does not explicitly introduce conflicting advocates. Here, the model is structured to capture the strength of opposing views indirectly, in the form of public opinion influencing the decisions of the regulator, the DM. However, the model in Kim (2014a) is clearly important for the purpose of designing a mechanism for making less error-prone decisions on controversial issues. This is the subject of the next step. Moreover, the court system differs from the task of this study in assuming the existence of truth; the conclusion of Kim (2014a) depends on the assumption that defendants know the truth about their guilt or innocence and that the probability of collecting evidence varies accordingly. In this study, as a severe accident is a once-off phenomenon, the existence of a distribution of true values is not considered important.

References

Argenziano, Rossella, Sergei Severinov, and Francesco Squintani 2016. "Strategic information acquisition and transmission." *Am. Econ. J. Micro.*. 8 (3): 119–55.

Beniers, Klaas J, Robert Dur, and Otto H Swank 2002. "Sequential Advocacy."

Che, Yeon-Koo, and Navin Kartik 2009. "Opinions as incentives." *J. Polit. Econ.*. 117 (5): 815–860.

Crawford, Vincent, and Joel Sobel 1982. "Strategic Information Transmission." *Econometrica*. 50 (6): 1431–51.

Dewatripont, Mathias, and Jean Tirole 1999. "Advocates." *Journal of political economy*. 107 (1): 1–39.

Dur, Robert, and Otto H Swank 2005. "Producing and manipulating information." *Econ. J.*. 115 (500): 185–199.

Gerardi, Dino, and Leeat Yariv 2008. "Costly expertise." *Am. Econ. Rev.*. 98 (2): 187–93.

Kim, Chulyoung 2014a. "Adversarial and inquisitorial procedures with information acquisition." *The Journal of Law, Economics, & Organization*. 30 (4): 767–803.

Kim, Chulyoung 2014b. “Partisan advocates.” *Bulletin of Economic Research*. 66 (4): 313–332.
Sobel, Joel 2013. “Giving and receiving advice.” *Advances in economics and econometrics*. 1 305–341.
Van den Steen, Eric 2010. “Culture clash: The costs and benefits of homogeneity.” *Manage. Sci.*. 56 (10): 1718–1738.

Chapter 3
Model Construction

3.1 Risky Project

Consider a decision to stop or execute a risky project that may cause severe damage to society. This society comprises a sufficiently large number of individuals. One of these individuals, referred to as the DM, has the authority to make the decision to either stop or execute the project. The execution of the project is delegated to an agent who is a member of society. For notational convenience, denote DM as "she" and the agent as "him."

The project is implemented over two periods. During the first period, research of the project will be conducted, and during the second period, the project may be in full swing. At the end of the first period, the DM decides whether the project will go into operation in the second period. If the decision is made to put the project into operation, it cannot be stopped through the second period. When the project goes into operation, it generates a fixed amount of benefit B $\left(B \in \mathbb{R}^{++}\right)$ to society in the second period, where $\mathbb{R}^{++} \equiv \{x \in \mathbb{R} | x > 0\}$. No benefits accrue from the first period of research. The benefit B generated in the second period is a public good and is assumed to accrue equally to all the members of society. In addition, the agent accrues a fixed value of private profit R $\left(R \in \mathbb{R}^{+}\right)$ from the operation of the project that is not a transfer from the society, where $\mathbb{R}^{+} \equiv \{x \in \mathbb{R} | x \geq 0\}$. In the first period, all the members of the society are aware of the existence of the benefit B and the private profit R, and correctly predict both values. The benefit to society from the alternative action taken when the project is not conducted (i.e., the outside option) is zero.

It is possible that an accident may occur if this project is operated in the second period. All members of society are aware of the possibility of the accident. In the first period, an event called a precursor occurs due to the investigation trial described later, but no accident occurs. If an accident occurs in the second period, it causes damage. This damage, like the benefit, is a public good and affects all members of society equally. The damage that an accident causes to society is denoted by a real value

A. Torii, *Decision-making in High-Risk Projects*,
https://doi.org/10.1007/978-981-97-3056-8_3

D $\left(D \in \mathbb{R}^{++}\right)$. The value of D is not known precisely by any member of society. Even if we observe this society from outside the system, we do not know whether the value of D follows a particular probability distribution. However, we assume each member of society has their own estimate of the damage and is confident in this estimate, reflecting their personal beliefs. In particular, the DM's belief of the damage is denoted as D^M. The accident does not affect the accrual of project benefits. In other words, if the project is carried out and an accident occurs during the second period, benefit B accrues and damage D occurs.

The mechanism by which the amount of damages in the event of an accident is determined as a stochastic event is not known. Meanwhile, the following is known about the probability of accident occurrence.

Assumption 3.1 1. The occurrence of accidents involves a random variable $X : \Omega \rightarrow [0, 1]$ defined in the probability space $(\Omega, \mathcal{F}, P)$, where Ω represents all possible outcomes of an observation; the σ-algebra $\mathcal{F}$ is a collection of subsets of Ω that we consider as "events" with associated probabilities, combined with a probability measure P, defined by $\mathcal{F}$. These elements form a probability space $(\Omega, \mathcal{F}, P)$.
2. The random variable X is known to follow a continuous uniform distribution $U(0, \theta)$. The value of the parameter θ is unique to the project and $\theta \in (0, 1]$.
3. In the first period of the project investigation, an event that is a precursor of an accident occurs with a probability $\pi(e)$ depending on the agent's level of effort e $(e \in \mathbb{R}^{+})$ when he performs an investigation trial. The precursor occurs as an event in the form that a value of the random variable X is observed, which is an element ω of the set Ω with probability corresponding to the level of effort e of the agent when he performs an investigation trial. An agent's survey attempt, regardless of whether it obtains a precursor, results in a cost of $c(e)$ to him depending on the level of effort. It is not possible for the other members of society to observe whether the agent has undertaken an investigation trial.
4. When the parameter θ satisfies $\theta > \tau$, where $\tau \in (0, 1)$, then if the project is executed in the second period, an accident occurs with probability 1 within the second period. However, when $\theta \leq \tau$, no accident occurs in the second period.

Here, an accident is assumed to occur with probability 1 or probability 0 in the second period, depending on the value of the parameter θ of the random variable X. This process can be understood as the continuous generation of the random variable X in the second period if the project is carried out; if X such that $X > \theta$ is once generated, the accident occurs. Since the second period is very long and the project cannot be stopped in the middle of the second period, if $\theta > \tau$, an event $\omega \in \Omega$ such that $X > \tau$ occurs with probability 1. This is equivalent to assuming the so-called Murphy's Law, or considering accidents as "normal accidents" in Perrow (1984, 1999). Meanwhile, when $\theta \leq \tau$, no accident occurs, because there is no event generated as such $X > \tau$.

This study explores scenarios in which an agent has the option to conduct up to two trials based on their decision. Within this probabilistic framework, we define the sample space Ω to encompass all possible outcomes of these trials. The observed random variables x_i are then introduced as specific realizations ω within Ω, with $i \in \{1, 2\}$ denoting the order of observations. This means each observation x_1 and x_2 corresponds to distinct outcomes in Ω, manifesting as vectors $\boldsymbol{x}_1$ and $\boldsymbol{x}_2$, respectively. For the first observation, the vector is represented as $(x_1)^t$, and for both observations combined, as $(x_1, x_2)^t$. In cases where no observations are obtained, we denote the outcome as $\boldsymbol{x}_0$, conceptualized as a zero-dimensional vector $\boldsymbol{x}_0(= \varnothing)$, within the specialized σ-algebra Σ^0, to represent the "no observation" scenario.

To systematically address the observation scenarios, we introduce Σ^0, Σ^1, and Σ^2 as subsets of the overarching σ-algebra $\mathcal{F}$, each tailored to specific contexts. Σ^0 captures the "no observation" scenario, while Σ^1 and Σ^2 correspond to spaces involving single and dual observations, delineated within intervals $[0, 1]$ and $[0, 1] \times [0, 1]$, respectively. Thus, Σ^0, Σ^1, and Σ^2 specify $\mathcal{F}$ by categorizing different observational contexts. Note that $\boldsymbol{x}_0 \in \Sigma^0$ is defined to represent the "no observation" scenario, illustrating the framework's ability to incorporate cases where no empirical data is obtained. While Σ^0 is specifically tailored to accommodate such scenarios, it is not itself an empty set but a σ-algebra that includes the concept of "no observation" as a meaningful element of analysis. Additionally, the notation $\boldsymbol{x}$ is used to denote a vector of observed values without specifying their count. The notation Σ represents the set of all possible events Σ^i for $i \in \{0, 1, 2\}$, encompassing observed values irrespective of their number. Henceforth, any observed value of the random variable X, serving as a precursor, will be referred to simply as "a precursor."

The information structure of this society is as follows. All members of the society do not know the value of the parameter θ of the random variable X specific to the project. However, the value of the real constant representing critical threshold τ is known exactly by all members of society. The DM's estimate of the accident damage, D^M, is also disclosed and is assumed to be known by all members of society. Furthermore, all members of society can observe the realized value of the precursor obtained by the investigation trial. In Chap. 6, this assumption is modified to consider the case in which only the agent that conducts the investigation trial in the first period can observe it and keep its value secret. As a precursor occurs only stochastically for an investigation attempt, members of society other than the agent know that the agent conducted the investigation trial when a precursor is obtained. However, when the random variable X is not obtained in the first period, members of society cannot distinguish whether the investigation trial was not conducted, or whether it was conducted but the precursor was not obtained. Furthermore, society members other than the agent cannot directly observe the agent's effort level e.

3.2 Decision-Making by the DM (1)

Consider the DM's decision. Assume the DM believes the project should be carried out as long as the benefits from the project do not fall below her expected value of the damage, considering the probability of accident occurrence. The benefits of the project are fixed at B by assumption. The DM is confident of D^M in terms of damages in the event of an accident, but the probability of occurrence can only be estimated. Here, if an investigation trial is conducted in the first period and the value of the random variable X can be observed as a precursor, the probability can be estimated based on the observed value. Thus, the DM is supposed to make the optimal decision for society using the information $\boldsymbol{x}$. For the decision, the following Assumption 3.2 is made.

Assumption 3.2 (*No commitment*) The DM makes the decision based only on given $\boldsymbol{x}$.

By Assumption 3.2, the DM does not consider factors other than given $\boldsymbol{x}$ in making the decision. In particular, the DM does not consider the possibility of implementing a contract that restricts the way in which the decision is made. As explained in Chap. 4, the way the decision is made affects the effort level of the agent in the investigating trial. Then, if the DM can commit to the way the decision is made by implementing a contract with the agent, the DM designs the optimal contract considering this effect. Then, the problem to be solved is not the optimal decision under the given institutional relationship between DM and the agent, but rather, is the optimal design of the contract. However, that is not the purpose of this study.

The DM, which would use the information of a precursor for decision-making, Bayesian updates her prior distribution of θ. At the start of the first period, the DM has a prior distribution of θ. Denote the posterior cumulative distribution function of parameter θ updated by the observed value $\boldsymbol{x}$ as $F(\theta|\boldsymbol{x})$ and the posterior density function as $f(\theta|\boldsymbol{x})$. Then, denote the probability derived from the posterior distribution function as $p(\cdot|\boldsymbol{x})$.[1] In particular, $p(\theta > \tau|\boldsymbol{x}) = 1 - F(\tau|\boldsymbol{x})$, that is the subjective probability of θ taking a value within the range $(\tau, 1]$, based on the distribution conditioned by the observed value $\boldsymbol{x}$.

The DM estimates the expected value of the difference between the benefit and the damage as

$$B - p(\theta > \tau|\boldsymbol{x}) \cdot D^M. \tag{3.1}$$

Then, the DM decides to start the project only when the value shown by the Eq. (3.1) is non-negative. This decision by the DM is equivalent to selecting subset $\sigma \subset \Sigma$, which point-wisely maximizes the function $W(\boldsymbol{x}; \sigma, B, D, \tau) : \Sigma \to \mathbb{R}$ for each $\boldsymbol{x}$ that is defined on the observed value $\boldsymbol{x} \in \Sigma$.

[1] The notation p represents the probability function, which is distinct from the probability measure P defined over the probability space.

Definition 3.1 (*Project value function of the DM*)

$$W(\boldsymbol{x}; \sigma, B, D^M, \tau) \equiv H_\sigma(\boldsymbol{x}) \cdot \left(B - p(\theta > \tau|\boldsymbol{x}) \cdot D^M\right), \tag{3.2}$$

where $H_\sigma(\boldsymbol{x})$ is the indicator function:

$$H_\sigma(\boldsymbol{x}) \equiv \begin{cases} 1 & \text{when } \boldsymbol{x} \in \sigma, \\ 0 & \text{when } \boldsymbol{x} \notin \sigma. \end{cases}$$

In this study, for convenience, the value of the indicator function on the empty set is specified as

$$H_\varnothing(\boldsymbol{x}_0) \equiv 1.$$

Although the empty set does not have elements, it is specified for consistency of expression. Moreover, "point-wisely maximize for each $\boldsymbol{x}$" refers to the maximization

$$\max_{\sigma \subset \Sigma} W(\boldsymbol{x}; \sigma, B, D, \tau) \;\; \forall \boldsymbol{x} \in \Sigma.$$

Obviously, this maximization is accomplished by setting σ such that $\boldsymbol{x} \in \sigma$ only when $B - p(\theta > \tau|\boldsymbol{x}) \cdot D^M \geq 0$. Subset σ represents the condition that the DM sets for the execution of the project. Since the only available information here is $\boldsymbol{x}$, the condition under which the DM sets whether to execute the project is shown by the subset of the set Σ of observed values. Only when the observed value is an element of the subset σ does the DM decides to execute the project.

Hereafter, the function (3.2) defined in Definition 3.1 is referred to as project value. The parameter D^M in the project value W is the estimate of the accident damage in the event of an accident, as believed by the entity evaluating the project value. Here, since the project value of the DM is being evaluated, the expected damage value is D^M. Later, the project value function in Definition 3.1 is extended to define the project value for entities other than the DM, and in that case, the value represents the belief of that entity. Parameter B is the benefit brought about by the project implementation. This value is assumed to be the same for all members of society; however, it is used here as a parameter, because this study later considers the case in which the private benefit R of the agent executing the project is also considered as an extension.

A condition σ^* is optimal if the subset σ^* maximizes the value of the function $W(\boldsymbol{x})$ point-wisely for each $\boldsymbol{x}$. In particular, the condition σ^* is defined as the optimal decision for the DM:

Definition 3.2 (*Optimal decision for the DM*)

$$\sigma^* = \{\sigma|\sigma \subset \Sigma,\ W(\boldsymbol{x};\sigma, B, D^M, \tau) \geq W(\boldsymbol{x};\sigma', B, D^M, \tau)\,,\ \forall \boldsymbol{x} \in \Sigma,\ \forall \sigma' \subset \Sigma\}.$$
$$\sigma_i^* = \{\sigma_i|\sigma_i \subset \Sigma^i,\ W(\boldsymbol{x};\sigma_i, B, D^M, \tau) \geq W(\boldsymbol{x};\sigma_i', B, D^M, \tau)\,,\ \forall \boldsymbol{x} \in \Sigma^i,\ \forall \sigma_i' \subset \Sigma^i\}.$$

When the number of observations is specified as $i,\ i \in \{0, 1, 2\}$, the optimal condition is represented as σ_i^*.

3.3 Decision-Making of the DM Under no Information

The prior distribution $F(\theta|\boldsymbol{x}_0)$ before or without obtaining a precursor is called the no-information prior distribution, for which a continuous uniform distribution is chosen here, as is usually assumed.

Assumption 3.3 (*no-information prior distribution*)

$$F(\theta|\boldsymbol{x}_0) = \begin{cases} 0 & \text{for } \theta \leq 0 \\ \theta & \text{for } 0 < \theta \leq 1 \\ 1 & \text{for } 1 < \theta \end{cases} \tag{3.3}$$

The probability density function of this distribution is $f(\theta|\boldsymbol{x}_0)$.

The subjective probability of an accident occurring under the no-information prior distribution is

$$p(\theta > \tau|\boldsymbol{x}_0) = 1 - F(\tau|\boldsymbol{x}_0) = 1 - \tau. \tag{3.4}$$

Under no-information prior distribution, the value of the project value function of the DM is

$$H_\sigma(\boldsymbol{x}_0) \cdot \left(B - (1-\tau)D^M\right).$$

Consider the optimal decision σ^* under no-information prior distribution. In this study, the main focus of the analysis is on the decision-making of how to decide to stop a project based on the precursor information, and thus, it is assumed that the DM decides to execute the project as the "status quo" under the no-information prior distribution:

Assumption 3.4 (*Status Quo*)

$$B - p(\theta > \tau|\boldsymbol{x}_0) \cdot D^M = B - (1-\tau)D^M > 0,\quad D^M > B.$$

For the first half of this assumption to make sense, the second half is assumed: if $B \geq D^M$, the first half of the assumption always holds and project execution is always desirable, rendering the analysis meaningless.

When no precursor is given, Σ^0 is specifically designed to represent the "no observation" scenario, and thus it incorporates $\boldsymbol{x}_0$ to symbolize cases where no empirical data is obtained. It follows that the subset of Σ^0 encompasses the scenario represented by $\boldsymbol{x}_0$. Then, the specification $H_\varnothing(\boldsymbol{x}_0) = 1$ implies that $\sigma_0^* = \varnothing$ is the optimal decision for this case, which is consistent with Assumption 3.4.

3.4 Update of the Distribution of Parameter θ Based on the Observation of a Precursor

Consider the update of the probability distribution of the parameter θ by the observed value of the precursor. Let y_2 denote the maximum of two observations, defined as $y_2 \equiv \max(x_1, x_2)$. When there is only one observation, let y_1 denote a single observation, defined for convenience as $y_1 \equiv x_1$. The update of the probability distribution is obtained by Bayes' theorem.

Lemma 3.1 (Update of prior distribution and the subjective probability of an accident occurring based on posterior distribution) *With the no-information prior distribution $F(\theta|\boldsymbol{x}_0)$, when $n = \{1, 2\}$ observation(s) are obtained, the posterior distribution of parameter θ is*

$$F(\theta|\boldsymbol{x}_1) = \begin{cases} 0 & \text{for } \theta \leq x_1 \\ -\dfrac{\log(\theta/x_1)}{\log(x_1)} & \text{for } x_1 < \theta \leq 1 \\ 1 & \text{for } 1 < \theta, \end{cases} \tag{3.5}$$

for case $n = 1$, and

$$F(\theta|\boldsymbol{x}_2) = \begin{cases} 0 & \text{for } \theta \leq y_2 \\ \dfrac{{y_2}^{-1} - \theta^{-1}}{{y_2}^{-1} - 1} & \text{for } y_2 < \theta \leq 1 \\ 1 & \text{for } 1 < \theta, \end{cases} \tag{3.6}$$

for case $n = 2$. The subjective probability of an accident occurring is

$$p(\theta > \tau|\boldsymbol{x}_1) = \begin{cases} \dfrac{\log(\tau)}{\log(x_1)} & \text{for } 0 < x_1 \leq \tau \\ 1 & \text{for } \tau < x_1 \leq 1. \end{cases}$$

for case $n = 1$, and

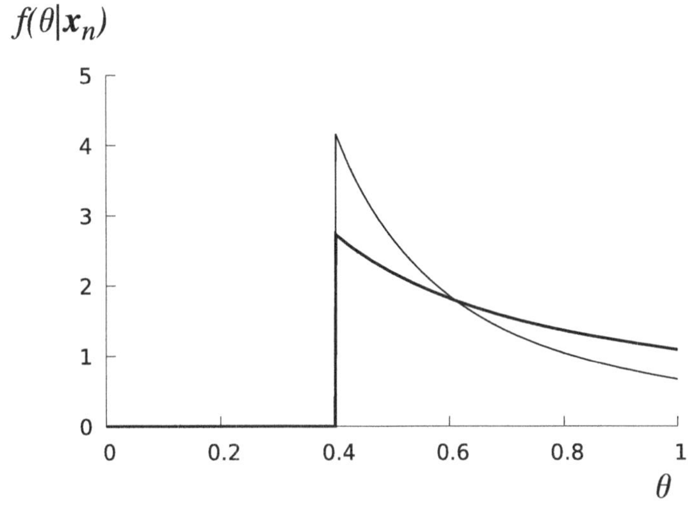

Fig. 3.1 Posterior distribution of parameter θ bold: $\boldsymbol{x}_1 = (0.4)$ thin: $\boldsymbol{x}_2 = (0.4, 0.4)^t$

$$p(\theta > \tau|\boldsymbol{x}_2) = \begin{cases} \dfrac{\tau^{-1} - 1}{y_2{}^{-1} - 1} & \text{for } y_2 < \theta \leq 1 \\ 1 & \text{for } 1 < \theta, \end{cases}$$

for case $n = 2$.

For example, Figs. 3.1 and 3.2 show the posterior distribution of the parameter θ when the observed values are $\boldsymbol{x}_1 = (0.4)$ and $\boldsymbol{x}_2 = (0.4, 0.4)^t$, respectively, including the subjective probabilities of an accident occurring based on the posterior distributions when $\tau = 0.8$.

3.5 Decision-Making by DM (2)

Consider decision-making by DM again. Given n precursor(s), the condition for executing the project for the DM is given by Lemma 3.2 with the definition x_n^*.

Definition 3.3 (x_n^*)

$$x_n^* \equiv \begin{cases} \tau^{D^M/B} & \text{for } n = 1, \\ \tau\delta & \text{for } n = 2, \end{cases} \quad \delta \equiv \left(\tau + \frac{D^M}{B}(1-\tau)\right)^{-1} < 1.$$

Lemma 3.2 (Decision-making with precursor) *Given random variable $\boldsymbol{x}_n$ as precursor, the optimal decision σ_n^* that maximizes point-wisely the project value for the DM, (3.2), is $\sigma_n^* = \{\boldsymbol{x}_n | y_n \in [0, x_n^*]\}$, $x_1^* < x_2^*$, and $\dfrac{\partial x_n^*}{\partial D^M} < 0$ for $D^M > 0$.*

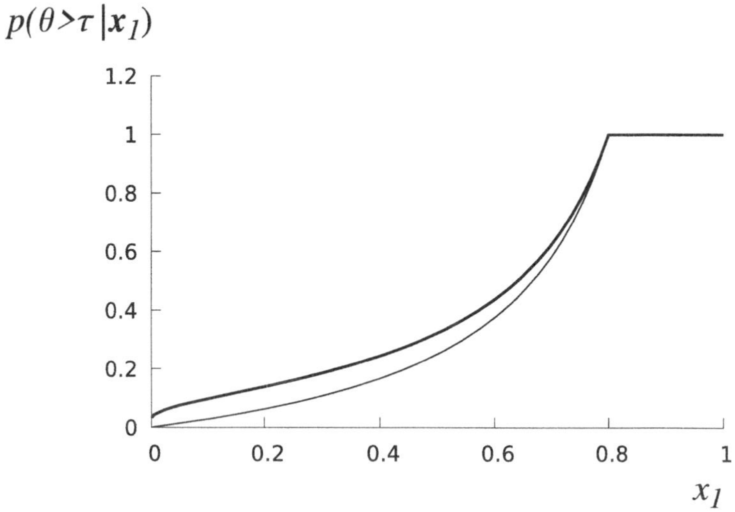

Fig. 3.2 Subjective probabilities of accident occurring based on Posterior distributions ($\tau = 0.8$) bold: $\boldsymbol{x}_1 = (0.4)$ thin: $\boldsymbol{x}_2 = (0.4, 0.4)^t$

According to Assumption 3.4, the project is presumed to be executed as the status quo in the absence of a precursor. A single observation as a precursor can overturn this decision, halting the project. Specifically, a value of θ exists that yields x_1 such that $x_1^* \leq x_1$. The acquisition of an additional observation may again overturn the decision, resulting in the project's execution. This indicates the existence of a value of θ that produces x_1 and x_2, satisfying the condition $x_1^* < x_1 \leq x_2 = y_2 < x_2^*$. Alternatively, even if one observation alone does not change the decision, a subsequent observation could still cause a reversal. In this scenario, a value of θ exists that produces x_1 and x_2 such that $x_1 < x_1^* < x_2^* < x_2 = y_2$. The condition $x_1^* < x_2^* < \tau$, implied by Lemma 3.2, suggests both scenarios.

If a condition σ is optimal, the project continues if and only if the project value for the DM is positive for the observed precursor. Here, the condition $B - p(\theta > \tau|x_1) \cdot D^M \geq 0$ can be taken as the condition indicating the acceptable range of subjective probability of the DM. The DM decides to proceed with the project only if the subjective probability of an accident is less than B/D^M. In other words, the optimal decision is to set the maximum risk acceptable for the DM at the probability B/D^M as the threshold.

3.6 Errors in Decision-Makings

DMs frequently face choices that must be made with incomplete information, inherently introducing the risk of errors. This potential for error can be bifurcated into two categories. The first category involves a decision to terminate a project that would have been advantageous to continue. The second category encompasses decisions to proceed with a project when termination would have been the more beneficial choice. If the DM's decision follows the criteria given by Lemma 3.2, the first type of error occurs when $x_n^* < y_n \leq \theta \leq \tau$, and the second type of error occurs when $y_n \leq x_n^* < \tau < \theta$. Following the terminology in hypothesis testing, the first type of error is a "Type I error" and the second type is a "Type II error." Consequently, the probabilities of committing Type I and II errors are denoted as γ_n and β_n, respectively:

Definition 3.4 (*Probability of Committing Errors*)

$$\gamma_n \equiv p(y_n > x_n^* | \theta \leq \tau, \boldsymbol{x}_0),\ \beta_n \equiv p(y_n \leq x_n^* | \theta > \tau, \boldsymbol{x}_0),\ n = 1, 2.$$

Each probability is given by Lemma 3.3.

Lemma 3.3 (Probability of Committing Errors, when $n = 1$)

$$\gamma_1 = 1 - \tau^{\frac{D^M}{B} - 1} \left\{ 1 - \left(\frac{D^M}{B} - 1 \right) \log(\tau) \right\}, \qquad \gamma_2 = (1 - \delta)^2,$$

$$\beta_1 = -\frac{\tau^{D^M/B} \log(\tau)}{1 - \tau}, \qquad \beta_2 = \tau \delta^2.$$

Figures 3.3, 3.4 and 3.5 show the probabilities of errors according to the value of τ when $D^M/B = 20$. Meanwhile, Figs. 3.6 and 3.7 show the probabilities of errors according to the value of D^M/B when $\tau = 0.95$. These figures represent the properties shown in Remark 3.1.

Remark 3.1 (*Magnitude of Errors*)

$$\lim_{\tau \to 1} \gamma_n = 0, \quad \lim_{D^M/B \to \infty} \gamma_n = 1, \quad \lim_{\tau \to 1} \beta_n = 1, \quad \lim_{D^M/B \to \infty} \beta_n = 0$$

Remark 3.1 illustrates that when the probability of risk occurrence is extremely low ($\tau \to 1$), the decision-making approach outlined in Lemma 3.2 leads to a significantly reduced likelihood of unnecessarily halting a project to avoid risk ($\gamma_n \to 0$). Conversely, there is a high probability of executing a project that should be stopped ($\beta_n \to 1$). Moreover, if the DM believes that the risk is high and aims to avoid it as much as possible ($D^M/B \to \infty$), the probability of stopping the project unnecessarily increases ($\gamma_n \to 1$). However, this approach significantly reduces the probability of executing a project that should be stopped ($\beta_n \to 0$).

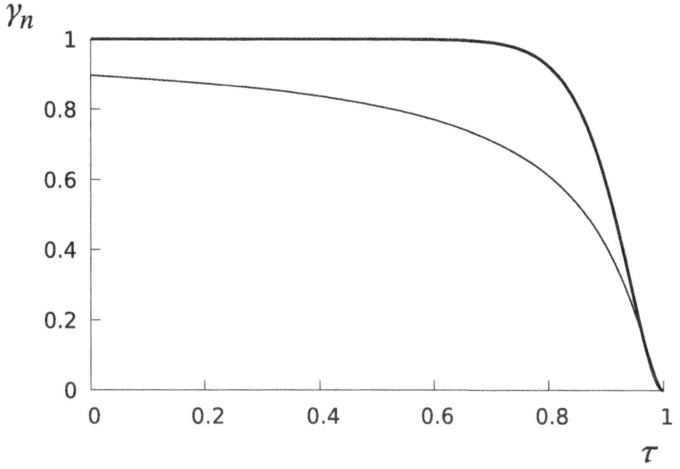

Fig. 3.3 Probability of Type I error (1) $(D^M/B = 20)$ bold: γ_1 thin: γ_2

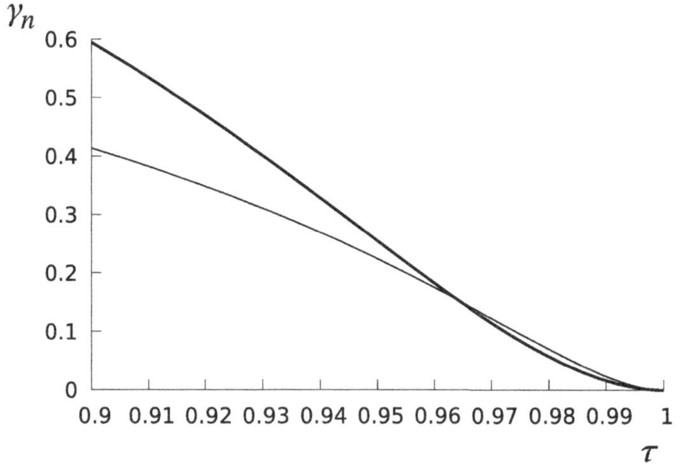

Fig. 3.4 Probability of Type I error (2) $(D^M/B = 20)$ bold: γ_1 thin: γ_2

The intersection of the graphs of γ_1 and γ_2, as well as β_1 and β_2, as depicted in the figures, suggests that under the decision-making framework of Lemma 3.2, having more observations does not necessarily reduce Type I and II errors.

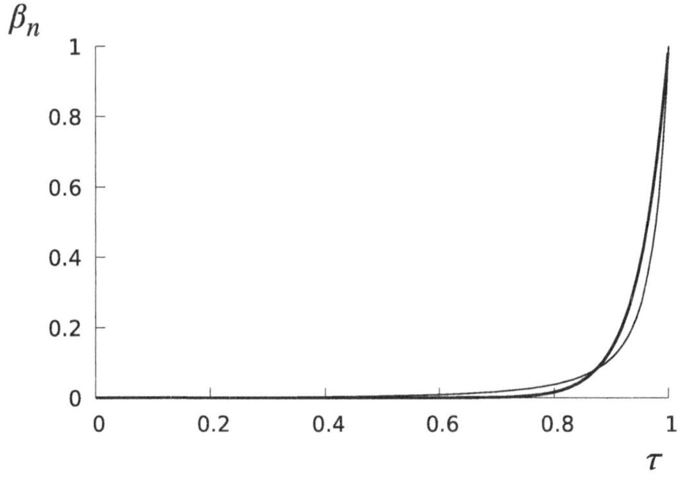

Fig. 3.5 Probability of Type II error (1) ($D^M/B = 20$) bold: β_1 thin: β_2

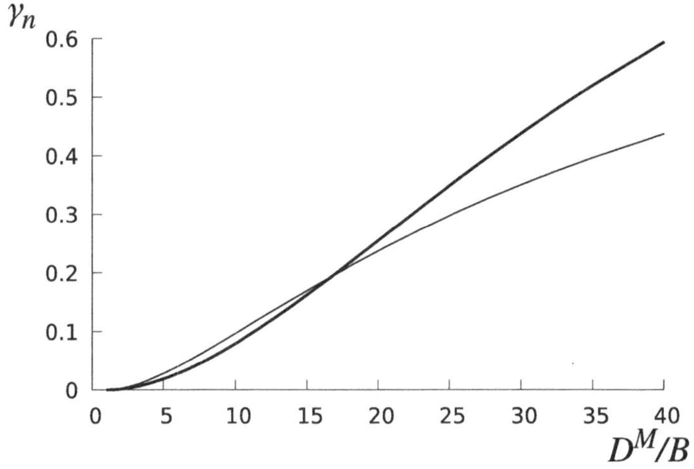

Fig. 3.6 Probability of Type I error (3) ($\tau = 0.95$) bold: γ_1 thin: γ_2

Furthermore, when considering the observed values as samples from a population, the decision-making process specified in Lemma 3.2 can be likened to a hypothesis test. Here, $\theta \geq \tau$ is the null hypothesis, and $\tau < \theta$ is the alternative hypothesis, with $(\sigma_n^*)^c = \{\boldsymbol{x}_n | y_n \notin [0, x_n^*]\}$ defining the rejection region. In this context, the Type I and II errors within the decision-making framework directly correspond to those in the hypothesis test.

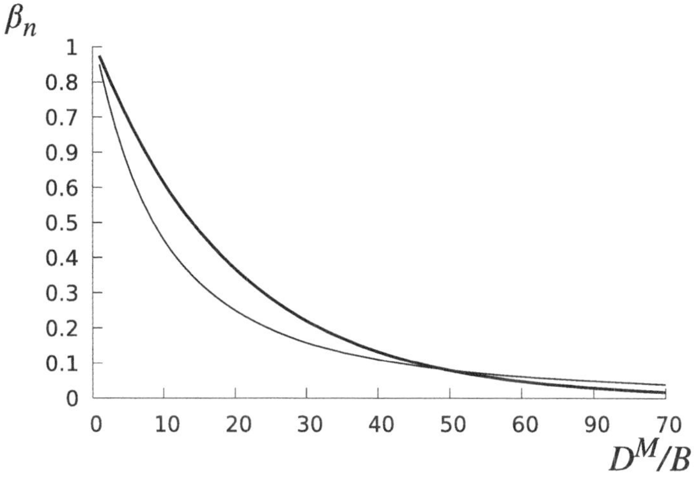

Fig. 3.7 Probability of Type II error (2) ($\tau = 0.95$) bold: β_1 thin: β_2

References

Perrow, Charles 1984. "Normal accidents: Living with high-risk technologies."

Perrow, Charles 1999. *Normal accidents: Living with high risk technologies.* : Princeton university press.

Chapter 4
Incentive to Acquire a Precursor

In this chapter, the incentive for the agent to obtain a precursor that is supposed to contribute to the DM's decision-making is analyzed. Only one investigation trial is assumed to be allowed. Chapter 5 presents an analysis where two trials are possible. For the trial, the agent undertakes effort that costs him, and for the effect of the effort, the following characteristics are assumed:

Assumption 4.1 (*Only One Investigation Trial Attempt*) $\pi(e)$ and $c(e)$ are twice-differentiable continuous functions.

$\pi(0) = 0, 0 \leq \pi(e) < 1, \pi'(e) \geq 0, \pi'(0) = \infty, \pi''(e) \leq 0$, for $e > 0$.

$c(0) = 0, 0 \leq c'(e), c'(0) = 0, 0 < c''(e)$, for $e > 0$.

$\exists\, \bar{e}, \bar{c} \in \mathbb{R}^{++}, c(\bar{e}) < \bar{c} < B, \forall e \geq \bar{e},\ \pi'(e) = 0$.

The agent can increase the probability of obtaining a precursor depending on the magnitude of the effort. Under Assumption 4.1, the agent has the option to forego the effort to obtain a precursor at all by selecting effort level $e = 0$. Effectively, this is equivalent to making no trial attempt. This distinction is inconsequential, as members of the society, other than the agent, are not presumed to have the ability to discern the presence or absence of an investigation trial. The assumption about $\bar{e}$ sets an upper bound on the level of effort. This also creates an upper bound $\bar{c}$ on the cost of effort. The upper cost limit is assumed to be less than or equal to the project's benefit B. These assumptions are made for the sake of simplicity in the analysis.

4.1 Project Value and Payoff for the Agent

D^A is defined as the agent's estimate of damages in the event of an accident. If the project proceeds into the second period, the agent, as a societal member, receives the project's benefits, B, along with potential damages in the event of an accident. By undertaking the project, the agent accrues a private profit, R, whose value is constrained as per the following assumption:

A. Torii, *Decision-making in High-Risk Projects*,
https://doi.org/10.1007/978-981-97-3056-8_4

Assumption 4.2 *(Private benefit)*

$$B > R > \bar{c}.$$

The private profit is a fixed amount that covers the cost of the effort. Then, the agent is not dissuaded from conducting the investigation trial due to financial constraints. The project's value for this agent is defined similarly to function (3.2) in Definition 3.1 for the DM, considering the agent's subjective probability of an accident:

Definition 4.1 *(Project value for the agent)*

$$W(\boldsymbol{x}; \sigma, B + R, D^A, \tau) = H_\sigma(\boldsymbol{x}) \cdot (B + R - p(\theta > \tau|\boldsymbol{x}) \cdot D^A) \tag{4.1}$$

When the DM decides to carry out the project, the agent receives the project benefit B as a public good, and the agent holds a belief of potentially suffering damage, amounting to D^A in the event of an accident, just like any other member of society. Even if the project's value is negative for the agent, he lacks the autonomy to terminate the project. This implies that as a societal member, the agent is invariably affected by the project and cannot avoid its impacts. For the agent, considerations regarding the participation constraint are confined to private profit and effort cost. Under Assumption 4.2, this participation constraint is always satisfied in this context.

It is assumed that the agent also shares the no-information prior distribution (Assumption 3.3) with the DM. Therefore, the posterior distribution of the parameter θ inclusive of the updated subjective probability is identical for both.

Consider the project value for the agent under the no-information prior distribution. Assuming the project is executed, the project value for the agent, based on the subjective probability of accident occurrence with a no-information prior distribution (3.4) is represented by W_0^A, as shown in the following definition:

Definition 4.2 (W_0^A, W_0^M)

$$W_0^A \equiv W_0^A(R, D^A, \tau) \equiv W(\boldsymbol{x}_0; \sigma_0^*, B + R, D^A) = B + R - (1 - \tau)D^A$$
$$W_0^M \equiv W_0^M(D^M, \tau) \equiv W(\boldsymbol{x}_0; \sigma_0^*, B, D^M) = B - (1 - \tau)D^M.$$

Note that $\sigma_0^* = \varnothing$. For subsequent reference, the project value for the DM, based on subjective probability within the same no-information prior distribution, is defined as W_0^M. To simplify notation and avoid typographical complications, $W_0^A(R, D^A, \tau)$ and $W_0^M(D^M, \tau)$ may be abbreviated as W_0^A and W_0^M, respectively.

The agent acquires one precursor, x_1, through an investigation trial with a probability, $\pi(e)$, that depends on the effort level, e. Upon observing a precursor, the DM obtains the posterior distribution $F(\theta|x_1)$ and the subjective probability $p(\theta > \tau|x_1)$ of an accident occurring, which informs the decision on whether to proceed with or halt the project. The agent knows the DM's decision-making scheme, as outlined by σ_1^* in Lemmas 3.1 and 3.2, and accordingly formulates his own project value. This project value for the agent, contingent upon the observation of a precursor x_1, is denoted as $W_1^a(x_1)$ and defined as follows:

Definition 4.3 $(W_1^a(x_1))$

$$W_1^a(x_1) \equiv W_1^a(x_1; R, D^A, \tau) \equiv W(x_1; \sigma_1^*, B + R, D^A),$$

where R, D^A, and τ are parameters. The project value is simply denoted as $W_1^a(x_1)$ to avoid notational complexity. The effort level is chosen to maximize the agent's own expected project value. At the point prior to the investigation trial, the expected project value, based on the subjective probability of observing one precursor, is represented by W_1^A, which is defined below.

Definition 4.4 (W_1^A)

$$\begin{aligned} W_1^A &\equiv W_1^A(R, D^M, D^A, \tau) \equiv E_{x_1}(W_1^a(x_1)|\boldsymbol{x}_0) \\ &= E_{x_1}(W(x_1; \sigma_1^*, B + R, D^A)|\boldsymbol{x}_0). \end{aligned}$$

In similar notations, the expression for the project value is simplified to W_1^A, representing a function that varies with parameters, rather than a constant. The expectation operator in this definition refers to the expectation regarding x_1's probability distribution, based on the subjective probability $F(\theta|\boldsymbol{x}_0)$. Lemma 4.1 details the exact values of W_1^a and W_1^A.

Lemma 4.1 (Project value for the agent based on one observed precursor)

$$W_1^a(x_1) = \begin{cases} B + R - D^A \dfrac{\log(\tau)}{\log(x_1)} & \text{for } 0 \le x_1 \le x_1^*, \\ 0 & \text{otherwise,} \end{cases}$$
$$W_1^A = x_1^* \left\{ D^A \log(\tau) - (B + R) \left(\log(x_1^*) - 1 \right) \right\}.$$

When the agent's expected accidental damage is greater than or approximately equal to that of the DM, the project value for the agent will increase, directly attributable to the observed precursor. This is shown by Lemma 4.2 and Definition 4.5.

Definition 4.5 $(\bar{D}^A)$

$$\bar{D}^A \equiv (B + R) \frac{1 - x_1^* + x_1^* \log(x_1^*)}{1 - \tau + x_1^* \log(\tau)}.$$

Lemma 4.2 (Increase of project value by observed precursor) *Under Assumptions 3.1, 3.2, 3.3, 3.4, and 4.2,*

$$\frac{\exp(1)-2}{\exp(1)-1}\frac{B+R}{B}D^M < \bar{D}^A < \frac{B+R}{B}D^M,$$

$$\textit{and} \quad \bar{D}^A \lesseqgtr D^A \iff W_0^A \lesseqgtr W_1^A \quad \forall D^M \in \mathbb{R}^{++}.$$

The numerical value of $(\exp(1)-2)/(\exp(1)-1)$ is around 0.42. Thus, unless the expected accident damage of the agent is not much lower than that of the DM, or his expected damage is higher than hers after adjusting for the effect of his private profits, the agent has an incentive to try to obtain a precursor.

Meanwhile, as long as the agent's private profit is positive, there is always a region in which he has no incentive to exert an effort. This region expands with the value of R. This can be easily checked as the threshold $\bar{D}^A$ increases as R increases. In this sense, the presence of larger private profits tends to hinder the efforts of the agent. This is reaffirmed in Proposition 4.1, which states that as private profits increase, efforts decrease correspondingly.

4.2 Determination of Effort to Acquire Information by Agent

The determination of the level of effort by the agent is analyzed here. When an agent expends effort e at cost $c(e)$, a precursor can be obtained with probability $\pi(e)$. The project value for the agent, determined probabilistically as W_1^A or W_0^A, depends on whether the observed value is present. Then, the agent's payoff is given by the expected project value minus the effort cost:

$$\pi(e) \cdot W_1^A + (1-\pi(e)) \cdot W_0^A - c(e) \tag{4.2}$$

Let e^* denote the effort level chosen by the agent to maximize his payoff.

Definition 4.6 (*Effort level chosen*)

$$e^* \equiv \underset{e}{\operatorname{argmax}}\{\pi(e) \cdot W_1^A + (1-\pi(e)) \cdot W_0^A - c(e)\}$$

Obviously, when $W_1^A - W_0^A > 0$, e^* satisfies the first-order condition:

$$\pi'(e^*) \cdot \left(W_1^A - W_0^A\right) - c'(e^*) = 0, \tag{4.3}$$

and when $W_1^A - W_0^A \leq 0$, $e^* = 0$.

From this condition, the next proposition about the effort level of the agent is proved.

Proposition 4.1 (Determination of Agent's Effort) *Under Assumptions 3.1, 3.2, 3.3, 3.4, 4.1, and 4.2,*

$$\bar{e} > e^* > 0, \quad \frac{\partial e^*}{\partial D^A} > 0, \quad \frac{\partial e^*}{\partial R} < 0$$

$\forall D^M, \forall D^A$, *such that* $D^A > \bar{D}^A$.

When the agent's own damage belief D^A is not much less than D^M, the agent's effort level is larger for larger D^A and smaller for smaller private profits R. Since the payoff for the agent is determined by the expected project value, it is natural that the greater the expected value of accident damages, the stronger the incentive to obtain a precursor and avoid the accident. Furthermore, given the assumptions in this context, an increase in private profit tends to diminish the incentive. This decrease arises because the project proceeds even without the observed value, and procuring a precursor invariably introduces the risk of halting the project.

Consider a situation where the agent's belief about accidental damages remains constant, while the DM's belief takes various values, being either greater or less than that of the agent. In this scenario, the agent's effort level reaches its maximum when the DM's belief is less than the agent's by an amount proportional to the ratio of private profit to the project's benefit, R/B. Proposition 4.2 further elucidates this relationship.

Proposition 4.2 (The relation between agent's effort and D^M) *Under Assumptions 3.1, 3.2, 3.3, 3.4, 4.1, and 4.2, when* D^A *is fixed,* e^* *is maximized at* $D^M = D^A B/(B + R) < D^A$.

When the agent's belief about accidental damage is fixed, there exists a peak in the DM's belief of damage, which is lower than the agent's belief. If the DM's belief is below this peak, the agent's incentive to exert effort increases with the DM's belief. However, if the DM's belief surpasses this peak, the agent's incentive to exert effort diminishes. The agent's criterion deviates from the DM's criterion by an amount equivalent to the private profit. If the private interest equals zero ($R = 0$), the decision-making criteria of both entities align when their beliefs about accident damage are identical. When the decision criteria coincide, the value of the information obtained from the agent's investigation trial is maximized. This is because if the decision criteria differ, the DM might not alter her decision based on new information, even if the agent believes that the additional information should influence the decision. In such a case, the impact of this additional information is diminished. Therefore, when the decision criteria of the DM and the agent diverge, the value of the precursor as a decision-making tool, along with the agent's incentive to acquire it, decreases. The value of the agent's investigation attempt reaches its maximum when the decision criteria of both entities coincide. However, when positive private interests exist, even if the beliefs of accident damage are identical, the decision criteria of the two parties are misaligned. At a point at which the DM's belief of accident damage is lower than the agent's, the decision criteria are effectively equal. At this point, the value of the information as a precursor is maximized, and thus, the effort incentive is also maximized. When the DM's belief of accidental damage is greater than this value, its value as information decreases, and, therefore, the agent's effort

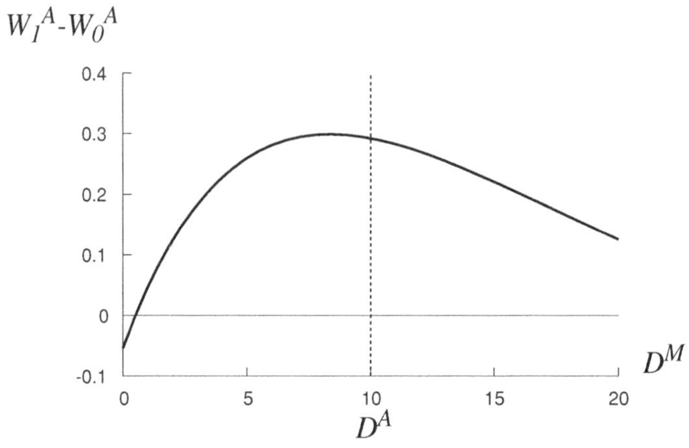

Fig. 4.1 Incentive $W_1^A - W_0^A$ for the agent as a function of D^M (when $D^A = 10$, $\tau = 0.9$, $B = 1$, $R = 0.2$)

incentive also decreases. For example, Fig. 4.1 shows how the incentive $W_1^A - W_0^A$ shapes as a function of D^M when D^A is fixed at a value of 10 (other parameters are set as $\tau = 0.9$, $B = 1$, and $R = 0.2$).

Proposition 4.2 can be interpreted as follows. In interpreting Lemma 3.2, the DM's decision-making can be understood as the same as setting the risk tolerance threshold at B/D^M and requiring that the probability of accident occurring be less than this threshold to execute the project. The same situation is assumed here, in that it is the agent that executes the project and the agent has his own expected project value and payoff from his subjective expected damage. Here, it is assumed that the agent is responsible for making decisions regarding the execution of the project, and that the acceptable risk threshold, B/D^M, on which the decision should be based is given externally as a regulation. Other members of society can also observe the precursor, but because of the high degree of specialization in both the investigation trial and execution of the project, recognizing and interpreting the precursor are left to the agent. Under such a setting, Proposition 4.2 implies that if the external regulation requires a lower and stricter standard of decision-making than the acceptable threshold of risk derived from the agent's own accident damage estimates, his effort incentive to acquire information is instead reduced. If a member of the society other than the agent anticipates that the accident causes very severe damage, and, therefore, imposes a stricter criterion for judgment, his level of effort to obtain a precursor is lower, the likelihood of obtaining a precursor is lower, and consequently, the probability of the project being carried out is higher.

Proposition 4.2 further suggests that if an agent, who holds an optimistic belief about the extent of accident damage, successfully captures the DM, who holds a more pessimistic belief than the agent, and aligns the decision-making with the agent's criteria, the agent's effort level actually increases due to the ability to influence this DM. That is, in the domain where $D^A < D^M$, the agent's effort level is a decreasing function of D^M. Therefore, when the value of D^M is redefined as D^A, the agent's effort level will invariably increase. By manipulating the DM's judgment, the final decision criterion aligns more closely with the agent's criterion, and as the value of additional information increases, the incentive to obtain that additional information increases. Therefore, the proposition that sufficient risk-assessment activities are not conducted as a consequence of the agent capturing the DM is not confirmed within this analysis framework.

In the state where the agent has captured the DM, the two entities are essentially the same as being conjoined. That is, if the DM also serves as an agent, executes projects, and makes decisions autonomously, a higher expected damage of both the DM and the agent results in a higher level of effort to acquire risk, as shown in the following Proposition 4.3.

Proposition 4.3 (Integration of DM and agent) *Under Assumptions 3.1, 3.2, 3.3, 3.4, 4.1, and 4.2, when the* DM *and the agent are the same, if* $e^* > 0$ *under a value of* $D^M = D^A$ *and* R *is not so large,* e^* *increases as the value of* $D^M = D^A$.

When there is an entity that both makes decisions and executes the project and the private benefit for the entity is not so large, the level of effort expended is the higher if the entity with the higher estimate of accident, damage is responsible for the project. This proposition implies that, when various entities are capable of making and carrying out decisions on their own, the project should be entrusted to an entity that is as pessimistic as possible about the amount of damages in the event of an accident.

4.3 Social Optimality

This study assumes that a severe accident involving a project can occur only once and that the probabilistic distribution of the damage amount is unknown. Even under these conditions, it is feasible to hypothesize a specific distribution of damages. However, using this assumed distribution to analyze social surplus optimality is inappropriate. We consider the decision-making problem under the assumption that there is no social consensus on the extent of damages caused by accidents, with each society member holding a unique expected damage value. Given a specific distribution, it is possible to contemplate the most efficient method for estimating the amount of accidental damages under certain assumptions. Consequently, we are left to evaluate the consequences of the DM's or the agent's behavior compared with the optimal one. However, when considering the issue of a once-off severe accident, such an analysis is not likely to be of great significance. We can never know whether the prediction of accidental damage is excessively low or excessively high, even after the occurrence of the accident.

Rather, recognizing that many real-life high-risk projects, as demonstrated in Chap. 1, may not have undergone sufficient risk-assessment efforts, this study questions why the level of risk-assessment effort is low. This study's aim and task are to analyze the determination of the effort level. Therefore, in this study, the levels of effort expended in the investigation trial are analyzed. Conducting deeper investigations into the probability of damage occurrence is considered important, especially when little is known about the distribution of damages. However, some propositions can also be made about social optimality, though in a different form than standard surplus analysis.

Assume that society members, including the DM, evaluate project values in the same way as the agent does. This is because these members of society are entities that receive equal benefits and suffer equal accidental damages from the project as public goods. If a member of society has a unique expected accidental damage value D, the project value for that member of society equals the project value for the agent W_0^A, W_1^A with $D^A = D$. These values are defined as $W_0(D)$ and $W_1(D)$, functions of the expected accident loss value, D, respectively.

Definition 4.7 (*Project value for society members*)

$$W_0(D) \equiv W_0(D; B, \tau) \equiv W(\boldsymbol{x}_0; \sigma_0^*, B, D, \tau) = B - (1-\tau)D, \tag{4.4}$$

$$\begin{aligned} W_1(D) &\equiv W_1(D; B, D^M, \tau) \equiv E_{x_1}(W(x_1; \sigma_1^*, B, D)|\boldsymbol{x}_0) \\ &= x_1^* \left\{B - (D^M - D)\log(\tau)\right\} \end{aligned} \tag{4.5}$$

As in the case of the agent, the project value is a random variable that depends on the observed values of a precursor. The expected value of this random variable is the payoff of these society members. However, as explained in the case of the agent, the payoff is still based on the expected accidental damage that each society member is subjectively convinced of. The level of this payoff is denoted as $Y(D, e^*)$, where

Definition 4.8 (*Payoff for society members*)

$$Y(D, e^*) \equiv \pi(e^*)W_1(D) + \left(1 - \pi(e^*)\right) W_0(D). \tag{4.6}$$

By Lemma 4.2, as long as $D > \bar{D}$, the update of the DM's decision by obtaining an observed value of a precursor increases the project value for that society member ($W_0(D) < W_1(D)$), where $\bar{D}$ is defined as

Definition 4.9 ($\bar{D}$)

$$\bar{D} \equiv B\frac{1 - x^* + x^*\log(x^*)}{1 - \tau + x^*\log(\tau)} < \bar{D}^A < D^M\frac{B+R}{B}.$$

Then, for the society members with a value of D that satisfies $D > \bar{D}$, their payoff increases with the agent's effort level, because

$$\frac{\partial}{\partial e^*} Y(D, e^*) = \pi'(e^*) \cdot (W_1(D) - W_0(D)) > 0.$$

Thus, Remark 4.1 is given.

Remark 4.1 (*Payoff of society members*) The payoff of a member of the society defined in Definition 4.8 is an increasing function of D^A if $D > \bar{D}$ and $D^A > \bar{D}^A$, and maximized at $D^M = D^A B/(B + R)$ when D^A is fixed. Furthermore, when the DM is the same entity as the agent, the payoff is larger if the entity with the larger expected loss is responsible for the project.

As a special case of Remark 4.1, if the society member in the remark is the DM, it increases its own payoff by delegating the investigation trial to an agent with higher estimates of accident damage. Conversely, the DM lowers her own payoff by delegating the investigation trial to a more optimistic agent with lower accident damage estimates.

4.4 Liability

In this section, the effect of imposing liability on the DM or the agent in the event of an accident is considered. Here, only the effect of setting a strict liability rule is considered. In the model, it is assumed that the agent's effort level e is unobservable, and thus, it is not possible to hold the agent liable for negligence by the reason that his effort is insufficient. As for the DM, the value of the random variable X as a precursor can be observed by all members, and the DM's own expected accident damage is assumed to be public information; thus, there is no discretion remaining to be blamed. Thus, this section investigates only the case of a strict liability rule that imposes liability in the event of an accident, regardless of the agent's level of effort and the appropriateness of the DM's judgment.

In general, if both the possible perpetrator and the possible victim could have avoided the accident by exercising due care, it would be desirable to have a negligence liability rule that encourages an optimum level of effort on both sides. Applying a strict liability rule that unilaterally imposes liability on either the perpetrator or the victim, regardless of the level of care, induces a moral hazard in which both parties try to save on the cost of care (Baird et al. 1998). However, if an economic activity of an economic entity is assumed to cause an accident, the strict liability rule is preferable because the damage caused by the accident is an external cost of the economic activity and can be internalized. Internalization minimizes the sum of the expected loss and the cost of care in an accident. The application of the negligence liability rule to accidents resulting from economic activities takes the form of a specific duty-of-care level, above which the exercise of care is exempt from liability. Under this liability system, the standard of care can be maintained at a desirable level, but the external effects caused by the accident of the economic activity in question are not internalized. It is argued that this leads to an excessive level of economic activity (Segerson 1995).

In this study, since an accident is assumed to cause extensive damage to society, imposing liability does not mean that a particular entity can compensate for all the damages. In the event of accidental damage, a portion of the amount of damages is imposed on the DM or the agent. Therefore, the effect of internalizing the external effects of the accident by imposing liability is also limited. However, the liability setting affects the level of effort the agent exerts to obtain a precursor. This study considers this effect on the level of effort to be important and includes it in the analysis.

First, consider the case in which strict liability is imposed on the agent. If an accident occurs during the execution of the project in the second period, a predetermined fixed amount of compensation is imposed on the agent, regardless of the investigation attempts made by the agent in the first period. Thus, strict liability works in the same way that the agent's expected accident damage D^A is increased by the amount of liability. This effect has already been analyzed in Proposition 4.1, which states that the agent's equilibrium effort level e^* is an increasing function of D^A. Thus, the following Remark 4.2 holds.

Remark 4.2 (*Strict liability imposed on the agent*) Under Assumptions 3.1, 3.2, 3.3, 3.4, 4.1, and 4.2, if $D^A > \bar{D}^A$, strict liability imposed on the agent increases e^*.

Imposing strict liability on the DM is more complicated. As in the case of imposing liability on the agent, in the DM's assessment of the positive or negative value of the project, she may consider the expected loss caused by the accident, which is the sum of the expected value of accidental damages and the liability amount. When society members observe this decision-making process of the DM, they may consider it as unjustified, because they may think the criteria should not be altered by the liability that is to be attributed to the DM's own responsibility. In such a case, where society considers the alteration of criteria by the amount of liability to be unjustified, the liability does not change any of the DM's decision-making. Meanwhile, if the society accepts a change in the decision, the decision changes in the form of an increase in the expected value of accident damages by the liability amount. Proposition 4.2 states that the agent's effort level is maximized at $D^M = D^A B/(B + R)$ when the DM's expected accident damage D^M changes, keeping the agent's expected accident damage constant. Thus, if the DM is more optimistic than the agent ($D^M < D^A$), imposing strict liability on the DM might increase the agent's effort level, although a large liability amount might actually undermine the agent's incentive to exert effort. Meanwhile, when the DM is more pessimistic than the agent ($D^M > D^A$), imposing strict liability on the DM impairs the agent's effort incentive. The following Remark 4.3 holds.

Remark 4.3 (*Strict liability imposed on the DM*) Under Assumptions 3.1, 3.2, 3.3, 3.4, 4.1, and 4.2, suppose that strict liability is imposed on the DM. If the DM's response to liability is to increase her own expected accidental damages by the amount of liability, the agent's chosen effort level e^* increases when $D^M < D^A B/(B+R)$ and the liability is small enough, and decreases when $D^M > D^A$.

Furthermore, when the DM and the agent are the same entity, by Proposition 4.3, imposing strict liability increases the incentive for the effort of that entity as stated in Remark 4.4.

Remark 4.4 (*Strict liability when the DM is the agent*) Under Assumptions 3.1, 3.2, 3.3, 3.4, 4.1, and 4.2, suppose that the DM and the agent are the same entity and that strict liability is imposed on the entity. If the response to the liability of the entity is to increase its own expected accidental damages by the amount of liability, the chosen effort level by the entity e^* increases as the value of $D^M = D^A$, as long as $e^* > 0$ under the value of $D^M = D^A$ and R is not so large.

References

Baird, Douglas G, Robert H Gertner, and Randal C Picker 1998. *Game theory and the law.* : Harvard University Press.

Segerson, Kathleen 1995."Liability and penalty structures in policy design." *Handbook of Environmental Economics, Basil Blackwell.*

Chapter 5
Incentive to Continue the Research Trial

5.1 Introduction

Continuing from the previous chapter, we examine the determination of the agent's incentive to seek information as precursors during project trials when the DM is making decisions in the initial period. It is postulated that up to two attempts to obtain a precursor via trials are possible. The decision to proceed with a second trial depends on the precursor obtained from the first trial. In some cases, despite the possibility of a second trial, the agent might choose not to exert any effort or seek additional precursors, depending on the outcome of the initial trial. The incentive to undertake further efforts for additional risk information, especially when some information has already been acquired, presents a different aspect from the incentive driven by the value of information from the trials, which is discussed in Chap. 4.

When multiple trials are permitted, the observed values from previous trials influence subsequent decision-making. Therefore, the decision regarding the amount of effort to invest in a trial must consider the possibility of discontinuing the trial based on its results. This consideration affects the incentive for effort in a trial. For a single trial, the incentive is determined by the magnitude of the effect on the expected change in project value given the precursor obtained from the trial. However, for multiple trials, the relative magnitude of the effect of information collection determines whether the trial continues, introducing a different factor.

To facilitate the analysis of trial continuation, this chapter introduces an assumption about the impact of the agent's effort on acquiring a precursor through trials. Specifically, with λ, $\bar{e}$, $\bar{c}$ as constants, we assume

Assumption 5.1 (Two Trials) $0 < c(e)$, $0 \leq c'(e)$, $0 < c''(e)$, for $e > 0$, $\lambda\bar{c} = c(\bar{e})$, $1 > \lambda > 0$,

$$\pi(e) = \begin{cases} 0 & \text{for } 0 \leq e < \bar{e} \\ \lambda & \text{for } e \geq \bar{e} \end{cases}$$

A. Torii, *Decision-making in High-Risk Projects*,
https://doi.org/10.1007/978-981-97-3056-8_5

Under this assumption, when the agent opts to exert effort during a trial, he incurs a cost of $\lambda\bar{c}$ and exerts an effort level of $\bar{e}$ because an effort level lower than $\bar{e}$ (i.e., $e < \bar{e}$) is less advantageous than exerting no effort at all ($e = 0$), and any effort greater than $\bar{e}$ (i.e., $e > \bar{e}$) is less effective than $\bar{e}$. By exerting an effort of $\bar{e}$, the agent can obtain observations as precursors with a constant probability λ, which is factored into their decision-making process.

Even under this setup, other members of society are unable to observe whether the agent genuinely exerts effort to gather risk-related information during the trial because the emergence of a precursor, indicative of the effort's result, is inherently uncertain. In this chapter, the analysis is based on Assumption 5.1, specifically focusing on whether the agent persists in their activities to acquire information.

5.2 Incentives for the Second Investigation Trial

The incentive for conducting a second investigation trial is contingent on whether a precursor was obtained through the first trial. To address this, we employ a backward-solving approach. Initially, we analyze the incentive for exerting effort in the second trial, assuming that a precursor has been acquired in the first trial. Consequently, the project value is determined based on the realized value of the random variable X, which represents the precursor from the first trial. The expected value of this project value then defines the incentive for exerting effort to obtain the initial precursor.

Consider the scenario where the agent has already obtained a single observational value $\boldsymbol{x}_1$ in the first trial. In this case, if the agent chooses to stop the trial with just this one observational value, it serves as a benchmark for decisions on whether to continue the trial. With only one observational value available, the DM decides to execute or halt the project based on this value, following the approach outlined in Lemma 3.2. Specifically, the project proceeds if $x_1 \leq x_1^*$; it halts if $x_1 > x_1^*$. In accordance with this decision-making method, the project's value to the agent, denoted as $W_1^a(x_1)$, is determined. At the point before obtaining the initial precursor, the expected value of this project's value, calculated with $\boldsymbol{x}_1$ distributed according to the no-information prior distribution, was represented by W_1^A.

Next, consider the change in the project's value when the agent continues their effort in the second trial. If the first observation, $\boldsymbol{x}_1$, is such that $x_1 > x_2^*$, the DM will not allow the execution of this project regardless of the presence of the second observation. Therefore, for the agent, the value of this project is 0. Consequently, no further effort will be made for the investigation trial, nor will any additional observation values be sought. However, if $x_1 \leq x_2^*$, there is a possibility that the project will be executed depending on the value of the second observation. Let the expected value of the project, at the point before conducting the trial to obtain the second observation, be denoted as $W_2^a(x_1)$, a function dependent on the possible second observation value, $\boldsymbol{x}_1$.

Definition 5.1 (W_2^a)

$$W_2^a(x_1) \equiv W_2^a(x_1; R, D^M, D^A, \tau) \equiv E_{x_2}\left(W\left(\boldsymbol{x}_2; \sigma_2^*, B+R, D^A\right)|\boldsymbol{x}_1\right)$$

Unless otherwise specified, $W_2^a(x_1; R, D^M, D^A, \tau)$ is abbreviated as $W_2^a(x_1)$.

We obtain Lemma 5.1 regarding the value of this $W_2^a(x_1)$.

Lemma 5.1

$$W_2^a(x_1) = \begin{cases} (B+R)\left(1 - \dfrac{1-\delta\tau+\log(\delta\tau)}{\log(x_1)}\right) + D^A\dfrac{\delta(1-\tau)}{\log(x_1)}, & \text{when } x_1 \in (0, x_2^*], \\ 0, & \text{otherwise.} \end{cases}$$

Under Definition 5.2, we can confirm Lemma 5.2 regarding the relation between $W_1^a(x_1)$ and $W_2^a(x_1)$. The value of τ^+ is approximately 0.6953.

Definition 5.2 (τ^+)

$$\frac{1}{1+\tau^+} + 2\log\left(\frac{\tau^+}{1+\tau^+}\right) - \frac{\log(\tau^+)}{1-\tau^+} = 0$$

Lemma 5.2 (Relation between $W_1^a(x_1)$ *and* $W_2^a(x_1)$) *Given* τ *such that* $\tau > \tau^+$, *under Assumptions 3.1, 3.2, 3.4,* and *4.2, there exists* D_+^M *and* D_+^A, *such that* $D_+^M \in (B, B/(1-\tau))$, $D_+^A \in (B, D^M)$, *and for all* $(D^M, D^A, \boldsymbol{x}_1)$ *such that* $D^M > D_+^M$, $D^A > D_+^A$, $\boldsymbol{x}_1 \in \{(x)|x \in (0, x_1^*]\}$,

$$W_2^a(x_1) - W_1^a(x_1) > 0, \quad \lim_{x_1 \to 0}\left\{W_2^a(x_1) - W_1^a(x_1)\right\} = 0,$$
$$\frac{dW_2^a(x_1)}{dx_1} - \frac{dW_1^a(x_1)}{dx_1} > 0.$$

Thus, if D^M is sufficiently large relative to B, and D^A is sufficiently close to D^M or $D^A > D^M$, the agent may try changing the project value from $W_1^a(x_1)$ to $W_2^a(x_1)$ by obtaining a second observation as a precursor with probability λ, by exerting effort at cost $\lambda\bar{c}$.

For the following convenience, we define D_+^M and $D_+^A(D^M)$ to identify the region of (D^M, D^A) where $W_2^a(x_1) > W_1^a(x_1)$ is guaranteed, as their existence has been proven in Lemma 5.2.

Definition 5.3 (D_+^M, D_+^A)

$$D_+^M \equiv \sup \left\{ D^M \left| \begin{array}{l} B\log(\tau\delta) - D^M \log(\tau) \\ +R(1-\delta\tau+\log(\delta\tau)) = 0, \\ \delta\left(\tau + \frac{D^M}{B}(1-\tau)\right) = 1, \\ D^M \in (B, B/(1-\tau)) \end{array} \right. \right\}$$

$$D_+^A(D^M) \equiv \inf\{D^A | W_2^a(x_1) > W_1^a(x_1)\}, \text{ for } D^M > D_+^M$$

As shown in the proof of Lemma 5.2, the value of $D_+^A(D^M)$ does not depend on $\boldsymbol{x}_1$.

As in the previous chapter, whether observations are made during the second trial, the project value, which is determined probabilistically, becomes a random variable. The payoff for the agent's effort during the second trial is calculated by subtracting the cost of effort from the expected increase in the project value. This payoff is based on the agent's subjective expectation of the accidental damage. Let this agent's payoff denoted as $Y_2^a(\boldsymbol{x}_1)$:

Definition 5.4 ($Y_2^a(\boldsymbol{x}_1)$)

$$Y_2^a(\boldsymbol{x}_1) \equiv \pi(e) \cdot W_2^a(x_1) + (1-\pi(e)) \cdot W_1^a(x_1) - c(e).$$

Under Assumption 5.1, the agent will exert effort $\bar{e}$ toward obtaining a second observation only if the condition

$$W_2^a(\boldsymbol{x}_1) - W_1^a(\boldsymbol{x}_1) \geq \bar{c} \tag{5.1}$$

is met. That is, considering the current precursor, the agent will undertake an additional trial only if it is expected that doing so will increase the project value by at least $\bar{c}$.

The region of x_1 where the agent exerts effort toward obtaining a second precursor is demonstrated in Lemma 5.3, under Definition 5.5.

Definition 5.5 (D_1^A, D_2^A, χ_1, χ_2,)

$$D_1^A \equiv \frac{B+R}{B} D^M, \quad D_2^A \equiv \frac{(B+R)D^M}{B}\left(1 + \frac{\log(x_2^*/x_1^*)}{1-x_2^*}\right).$$

Meanwhile, given a sufficiently small value of $\bar{c}$, χ_1 and χ_2 are defined as the values that satisfy the conditions

$$W_2^a(\chi_1) - \bar{c} = W_1^a(\chi_1), \quad \chi_1 \in (0, x_1^*),$$

and

$$W_2^a(\chi_2) - \bar{c} = 0, \quad \chi_2 \in (x_1^*, x_2^*),$$

respectively.

Lemma 5.3 (Regions where the agent exert effort) *Under Assumptions 3.1, 3.2, 3.3, 3.4, 4.2, and 5.1, given* D^M, D^A, *and* τ *such that* $D^M > D^M_+$, $D^A > D^A_+(D^M)$, *and* $\tau > \tau^+$, *when* $\bar{c}$ *is sufficiently small relative to B, the region of* $\boldsymbol{x}_1$ *where the condition that the agent exerts effort for the second observation as a precursor (as outlined in condition (5.1)) is satisfied is when* $D^A \in (D^A_+(D^M), D^A_1)$, $[\chi_1, x_2^*]$, $D^A \in [D^A_1, D^A_2)$, $[\chi_1, \chi_2]$, *and* $D^A \in [D^A_2, \infty)$, $[\chi_1, x_1^*]$.

According to Lemma 5.3, when the outcome of the agent's first survey results in the observed random variable X being within a specific range that depends on the value of D^A, the agent will conduct a second investigation trial. The probability, from the agent's perspective, of continuing to a second trial at the point before the first trial (under a no-information prior distribution) is described in Lemma 5.4.

Lemma 5.4 (Probability of continuing to a second trial) *Under Assumptions 3.1, 3.2, 3.3, 3.4, 4.2, and 5.1, when* $D^M > D^M_+$, $\tau > \tau^+$, *and* $\bar{c}$ *is sufficiently small compared to B,*

$$p(W_2^a(x_1) - W_1^a(x_1) \geq \bar{c}|\boldsymbol{x}_0) =$$
$$\begin{cases} -\displaystyle\int_{\chi_1}^{x_2^*} \log(x)\,dx & \text{when } D^A \in (D^A_+(D^M), D^A_1), \\ -\displaystyle\int_{\chi_1}^{\chi_2} \log(x)\,dx & \text{when } D^A \in [D^A_1, D^A_2), \\ -\displaystyle\int_{\chi_1}^{x_1^*} \log(x)\,dx & \text{when } D^A \in [D^A_2, \infty). \end{cases} \tag{5.2}$$

Regarding the subjective probability of continuing efforts until obtaining a precursor for the second time, Proposition 5.4 holds.

Proposition 5.4 (Effort to have the second precursor) *Under Assumptions 3.1, 3.2, 3.3, 3.4, 4.2, and 5.1, for all* D^M, D^A, *and* τ *such that* $D^M > D^M_+$, $D^A > D^A_+(D^M) > 0$, *and* $\tau > \tau^+$,

$$\frac{d\chi_1}{dD^A} < 0, \quad \lim_{\bar{c}\to 0} \chi_1 = 0.$$

Furthermore, for $D^A \in (D^A_+(D^M), D^A_1) \cup [D^A_2, \infty)$ *and sufficiently small* $\bar{c}$,

$$\frac{d}{dD_A} p(W_2^a(x_1) - W_1^a(x_1) \geq \bar{c}|\boldsymbol{x}_0) > 0.$$

The value of χ_1 represents a lower limit for the agent below which there is no need to attempt a second investigation trial to obtain another observation. According to Proposition 5.4, as the agent's accidental damage estimation increases, this lower limit decreases. Consequently, the agent is more inclined to conduct a second investigation, even for smaller values of the precursor. Thus, the incentive for further investigation increases. When $\bar{c}$ is sufficiently small, this limit approaches zero. Hence, if the cost of efforts to detect an early precursor is low enough, the agent has no incentive to restrict their efforts to just one attempt. These propositions are valid as long as $D^A > D^M$ or D^A is not significantly different from D^M. That is, even if the agent is optimistic and underestimates the amount of accident damage, this alone does not motivate him to limit his effort to detect early signs to just one attempt, provided the cost of doing so is sufficiently low.

The latter half of Proposition 5.4 demonstrates that the agent with higher estimated value of accident damage, D^A, is more likely to make an effort to continue obtaining a second precursor. This tendency is first confirmed in the region where $D^A \in (D^A_+(D^M), D^A_1)$. As $D^A_+(D^M) < D^M < D^A_1$, this situation arises when the agent's accident damage estimate is not significantly different from that of the DM. In such cases, as the agent's expected accident damage amount increases, he is more inclined to conduct a second investigation, even if the initial observation value x_1 is small, to avoid accidents. Meanwhile, he will consistently attempt to reconfirm up to an observation value of x_2^* to explore the feasibility of carrying out the project.

Additionally, this tendency is also evident in the region when $D^A \in [D^A_2, \infty)$. In these instances, the agent's expected accident damage is extremely high, and he shows a similar propensity to conduct a second investigation with even smaller observation values x_1 to prevent accidents. However, he tries to reconfirm the precursor by conducting a resurvey as long as the observation value does not exceed x_1^*. If the value exceeds x_1^*, he will abandon the project without waiting for the outcome of the second investigation attempt.

In the previous chapter, it was confirmed that agents with higher estimated accident damage values tend to exert a higher level of effort to obtain a (one-time) precursor. The same tendency is observed in the context of continued efforts in trials.

When the value of D^A is in the intermediate range, that is, $D^A \in [D^A_1, D^A_2)$, the tendency for the agent to conduct a second investigation with even smaller observation values x_1 to avoid accidents remains unchanged as the expected accident damage amount increases. However, there is a situation where, if a re-investigation is not conducted, the DM will abandon the project, though there is a possibility of project execution through re-investigation. In this context, the agent with higher expected accident damage value tends to let the DM abandon the project without conducting a re-investigation, even if the first observation value is smaller. Therefore, the range of observation values for which a re-investigation is conducted may shrink. Generally, since the former effect is stronger, agents with higher estimated accident damages tend to make more effort in re-investigation, similar to other regions. However, for example, if the cost of investigation effort is very small, the lower limit of the observation value for conducting a re-investigation almost sticks to zero and does not change, so only the latter effect occurs, leaving a possibility that agents with higher accident damage values are less likely to conduct a re-investigation.

5.3 Incentives for the First Investigation Trial

In the analysis in Sect. 5.2, the value of the project is presented for the case where an observation value of x_1 is obtained for the random variable X as a result of the first investigation trial. The decision to exert effort during the first trial is similar to the decision of whether to exert effort in the second trial. Effort is only exerted if the expected value of the project, by obtaining another precursor, increases by at least the probability-equivalent cost $\bar{c}$ of obtaining that precursor.

At the point of conducting the first investigation trial, consider the payoff of exerting effort $\bar{e}$. The project value can be obtained by integrating the project value for each case where the precursor obtained as a result of the trial is $\boldsymbol{x}_1$ with the probability density function (prior distribution) of $\boldsymbol{x}_1$. If a precursor is obtained and the effort in the trial is stopped after one occurrence, this expected value is represented by W_1^A. However, as it is possible to continue the effort in the second trial, the agent, upon conducting the first trial and obtaining the value of $\boldsymbol{x}_1$ as a precursor, can decide whether to stop the continuation of the trial and wait for the DM's decision based solely on the value of $\boldsymbol{x}_1$ or expend the cost $\lambda\bar{c}$ to continue the effort and hope to obtain another precursor. This decision is represented by comparing the project value if the effort in the trial is stopped, $W_1^a(x_1)$, with the expected value if the effort is continued, $W_1^a(x_1) + \lambda(W_2^a(x_1) - W_1^a(x_1) - \bar{c})$. It was demonstrated that the agent would choose the latter if $W_2^a(x_1) - W_1^a(x_1) - \bar{c} \geq 0$. The expected value of the project, which is determined based on the randomly obtained value of the precursor $\boldsymbol{x}_1$, is itself a random variable. This value, represented by $Y_1^a(\bar{e})$, is never less than the value of W_1^A, as shown in Lemma 5.5.

Definition 5.6 ($Y_1^a(\bar{e})$)

$$Y_1^a(\bar{e}) \equiv E_{x_1}\left(\max(W_1^a(x_1), W_1^a(x_1) + \lambda\left\{W_2^a(x_1) - W_1^a(x_1) - \bar{c}\right\})|\boldsymbol{x}_0\right)$$

Lemma 5.5

$$Y_1^a(\bar{e}) \geq W_1^A.$$

On the contrary, if the first trial opportunity is not utilized, the state of the project at the point just before the second trial becomes the same as if only one trial was allowed. If the project proceeds without obtaining a precursor and without using the opportunity to obtain a second precursor, the project value is W_0^A. Furthermore, if effort $\bar{e}$ is exerted in the trial to capitalize on the last opportunity to secure a precursor, the expected value of the project value is $W_0^A + \lambda\left(W_1^A - W_0^A - \bar{c}\right)$. From this, the following lemma is established.

Lemma 5.6 (Skipping the first trial) *In the situation where the agent chooses not to aim for a precursor with $e = 0$ during the first investigation trial but decides to exert an effort of $\bar{e}$ in a subsequent second investigation trial to obtain a precursor, it is necessary for the condition $W_1^A - W_0^A - \bar{c} \geq 0$ to be met.*

With the above preparations, the agent's decision on whether to exert positive effort to obtain a precursor during the first investigation trial can be demonstrated in the following Proposition 5.5.

Proposition 5.5 (Effort to obtain a precursor during the first trial) *Under Assumptions 3.1, 3.2, 3.3, 3.4, 4.2,* and *5.1, the necessary and sufficient condition for the agent to exert effort at the level of* $\bar{e}$ *to obtain a precursor during the first trial is that under the conditions* $D^M > D_+^M$, $D^A > D_+^A(D^M)$, *and* $\tau > \tau^+$, *it must hold that* $Y_1^a(\bar{e}) - W_0^A \geq \bar{c}$. *The region of* $\bar{c}$ (> 0) *where this condition is satisfied expands with increasing* D^A.

This proposition reveals that the agent with a higher expected accident damage is willing to exert more effort, up to a higher cost level, to obtain a precursor. In this sense, it becomes evident that the incentive for investigation trials among agents is greater for those who are more pessimistic.

Chapter 6
Confidentiality of Observed Precursor

6.1 Introduction

This chapter, unlike the previous chapters, considers a case in which the agent can keep secret the value of the stochastic variable X, the precursor. Up until now, the analysis has been based on the assumption that the value of X is observed by all members of the society as public information. This assumption is relaxed here, which considers the case in which only the agent observes the value of X when it is generated as private information in the course of the investigation trial, and the agent decides whether to report this value to the DM and other members of society. However, while the information can be kept secret, it cannot be falsified to manipulate its content.

As the agent and the DM are convinced of their own accident damage estimates, even if they do not disagree on the probability of an accident occurring due to shared observations of the precursor, they make different judgments on whether the project should be executed. Furthermore, the existence of private profit for the agent also causes divergent judgment. Each project value function is defined for the DM as shown in Definition 3.1, and for the Agent, it is defined as shown in Definition 4.1. The conditions for point-wise optimization for observed values $\boldsymbol{x}$ are:

$$B - p(\theta > \tau|\boldsymbol{x}) \cdot D^M \geq 0$$

for the DM, and

$$B + R - p(\theta > \tau|\boldsymbol{x}) \cdot D^A \geq 0$$

for the agent.

In the presence of such disagreement, the agent has an incentive, based on his own belief about the amount of accident damage, to pretend that he could not observe the precursor and to keep the information confidential. In this chapter, as an extension of the analysis in Chap. 4, the effect of information concealment possibility is analyzed.

A. Torii, *Decision-making in High-Risk Projects*,
https://doi.org/10.1007/978-981-97-3056-8_6

6.2 Case Information is Concealed

If positive effort in the investigation trial results in no observables, there is nothing different from the analysis in Chap. 4, since there is nothing to conceal. Consider the case in which the effort results in an observed value x_1. Let σ_1^A be the decision that optimizes the project value for the agent point-wisely for the observed value x_1. The DM's decision described in Sect. 3.5 is denoted by σ_1^*. Here, another decision is considered that would optimize the agent's own project value if he were able to make such a decision. This decision σ_1^A is shown in Lemma 6.1 with additional definitions for denotations.

Definition 6.1 (*Optimality for the agent*)

$$\sigma_1^A \equiv \Big\{\sigma | \sigma \subset \Sigma_1, W(\boldsymbol{x}; \sigma, B+R, D^A) \geq W(\boldsymbol{x}; \sigma', B+R, D^A), \\ \forall \boldsymbol{x} \in \Sigma_1, \forall \sigma' \subset \Sigma_1\Big\}.$$

Definition 6.2 (x_1^A)

$$x_1^A \equiv \tau^{\frac{D^A}{B+R}}.$$

Lemma 6.1 (Optimal decision for the agent)

$$\sigma_1^A = \{x_1 | x_1 \in [0, x_1^A]\}.$$

The agent keeps the observed values secret only when σ_1^* and σ_1^A are different. The discrepancy between σ_1^* and σ_1^A occurs in the forms shown in Lemma 6.2.

Lemma 6.2 (Discrepancy between the DM and the agent)

$$\begin{cases} \sigma_1^A \subset \sigma_1^*, \text{ when } \dfrac{D^A}{B+R} > \dfrac{D^M}{B}, \\ \\ \sigma_1^A \supseteq \sigma_1^*, \text{ when } \dfrac{D^A}{B+R} \leq \dfrac{D^M}{B}. \end{cases}$$

First, case $x_1 \in \sigma_1^A \cap (\sigma_1^*)^c$ is where the agent thinks the project should be executed but the DM thinks it should be stopped; second, case $x_1 \in (\sigma_1^A)^c \cap \sigma_1^*$ is where the agent thinks the project should be stopped but the DM thinks it should be executed.

When the agent's observations are concealed, there are two possibilities for the DM's decision-making: The DM may decide to proceed with the project or she may decide to halt execution. The following illustrates how each of the following four cases can be an equilibrium because of the two different judgments and the two different decisions.

Case1 $\sigma_1^A \supset \sigma_1^*$, $H_{\sigma_0^*}(\boldsymbol{x}_0) = 0$
Case2 $\sigma_1^A \subset \sigma_1^*$, $H_{\sigma_0^*}(\boldsymbol{x}_0) = 1$
Case3 $\sigma_1^A \supset \sigma_1^*$, $H_{\sigma_0^*}(\boldsymbol{x}_0) = 1$
Case4 $\sigma_1^A \subset \sigma_1^*$, $H_{\sigma_0^*}(\boldsymbol{x}_0) = 0$

In the previous chapters, the optimal decision σ_0^* was assumed as $H_{\sigma_0^*}(\boldsymbol{x}_0) = 1$, which is consistent with Assumption 3.4. This chapter also considers the case in which $H_{\sigma_0^*}(\boldsymbol{x}_0) = 0$ as the strategy for the DM; that is, the decision to halt the project is a no-information decision. This is because stopping project execution when there is no information works as an incentive for the agent to disclose information. This does not mean Assumption 3.4 is abandoned.

6.3 Expected Project Value for Each Case

Prior to the analysis, the expected values of the project value for the DM and the agent in each case are shown. For ease of representation, the following set is defined.

Definition 6.3 (ρ)

$$\rho \equiv \begin{cases} \sigma_1^* \cup \sigma_1^A & \text{when } H_{\sigma_0^*}(\boldsymbol{x}_0) = 1 \\ \\ \sigma_1^* \cap \sigma_1^A & \text{when } H_{\sigma_0^*}(\boldsymbol{x}_0) = 0 \end{cases}$$

When $H_{\sigma_0^*}(\boldsymbol{x}_0) = 1$, the scope of project execution always extends to the union set of σ_1^A and σ_1^*, regardless of the inclusion relation between σ_1^A and σ_1^*. In the case of $\sigma_1^* \subset \sigma_1^A$, if the value of the random variable X falls within the range $\sigma_1^A \cap (\sigma_1^*)^c$, the project is executed by concealing the information, while in the case of $\sigma_1^* \supset \sigma_1^A$, if the value of the random variable X falls within the range $\sigma_1^* \cap (\sigma_1^A)^c$, the project is executed regardless of whether it is concealed. Conversely, when $H_{\sigma_0^*}(\boldsymbol{x}_0) = 0$, the scope of project execution is always reduced to the product set of σ_1^A and σ_1^*. In the case of $\sigma_1^* \subset \sigma_1^A$, if the value of the random variable X falls within the range $\sigma_1^A \cap (\sigma_1^*)^c$, the project is stopped regardless of whether the observed value is concealed. Meanwhile, in the case of $\sigma_1^* \supset \sigma_1^A$, if the value of the random variable X falls in the range $\sigma_1^* \cap (\sigma_1^A)^c$, the project can be stopped by concealing information.

Thus, the decision ρ defines the extent to which project execution is effectively decided in each case when the precursor observation can be kept concealed. Based on this decision, ρ, the expected project value for the DM and the agent is expressed as follows:

Definition 6.4 (V_i^A, V_i^M) For each Case i ($i = 1, 2, 3, 4$),

$$V_i^A \equiv (1 - \pi(e)) \cdot W(\boldsymbol{x}_0; \sigma_0^*, B + R, D^A) + \pi(e) \cdot E_{x_1}\left(W(x_1; \rho, B + R, D^A)|\boldsymbol{x}_0\right),$$

$$V_i^M \equiv (1 - \pi(e)) \cdot W(\boldsymbol{x}_0; \sigma_0^*, B, D^M) + \pi(e) \cdot E_{x_1}\left(W(x_1; \rho, B, D^M)|\boldsymbol{x}_0\right),$$

respectively.

The expected project value of the agent and the DM in each case is shown in the following Lemma 6.3, where, for simplicity of representation, the following function is used.

Definition 6.5

$$\Phi(B', D', \sigma') \equiv \int_0^1 \int_{\sigma'} \{B' - p(\theta > \tau | x_1) \cdot D'\} \frac{dx_1}{\theta} dF(\theta | \boldsymbol{x}_0)$$

Lemma 6.3 (Expected project value for each case)

Case 1
$$V_1^A = \pi(e) \cdot \Phi(B + R, D^A, \sigma_1^*) = \pi(e) \cdot W_1^A$$
$$V_1^M = \pi(e) \cdot \Phi(B, D^M, \sigma_1^*)$$

Case 2
$$V_2^A = (1 - \pi(e)) \cdot W_0^A + \pi(e) \cdot \Phi(B + R, D^A, \sigma_1^*)$$
$$= (1 - \pi(e)) \cdot W_0^A + \pi(e) \cdot W_1^A$$
$$V_2^M = (1 - \pi(e)) \cdot W_0^M + \pi(e) \cdot \Phi(B, D^M, \sigma_1^*)$$

Case 3
$$V_3^A = (1 - \pi(e)) \cdot W_0^A + \pi(e) \cdot \Phi(B + R, D^A, \sigma_1^A)$$
$$V_3^M = (1 - \pi(e)) \cdot W_0^M + \pi(e) \cdot \Phi(B, D^M, \sigma_1^A)$$

Case 4
$$V_4^A = \pi(e) \cdot \Phi(B + R, D^A, \sigma_1^A)$$
$$V_4^M = \pi(e) \cdot \Phi(B, D^M, \sigma_1^A)$$

The function $\Phi(B', D', \sigma')$ has the following properties.

Lemma 6.4

$$\Phi(B + R, D^A, \sigma_1^A) = (B + R)\tau^{\frac{D^A}{B+R}} \geq \Phi(B + R, D^A, \sigma_1), \ \forall \sigma_1 \subset \Sigma^1,$$
$$\Phi(B, D^M, \sigma_1^*) = B\tau^{\frac{D^M}{B}} \geq \Phi(B, D^M, \sigma_1), \ \forall \sigma_1 \subset \Sigma^1.$$

For each case, the agent's effort level e is chosen to maximize the agent's payoff. e_i denotes the effort level selected for case i:

Definition 6.6 (e_i)

$$e_i \equiv \underset{e}{\text{argmax}} \left(V_i^A - c(e)\right) (i = 1, 2, 3, 4)$$

In the analysis of Chap. 4, when the value of the observed random variable X cannot be concealed, the chosen effort level is e^* given by the first-order condition (4.3). In the case in which the observed value can be kept concealed, the effort level e_i selected is equal to or greater than e^*, as shown in the following Proposition 6.1.

Proposition 6.1 (Effort level when information can be concealed) *Under Assumptions 3.1, 3.2, 3.3, 3.4, 4.1, and 4.2, $W_0^A \geq 0 \Rightarrow e_i \geq e^*$ when $D^A > \bar{D}_A$. The possibility of $e_i < e^*$ is limited to Cases 1 and 4. In particular, in Case 1, $W_0^A < 0 \Rightarrow e_1 < e^*$.*

When $W_0^A \geq 0$, only in Case 2, the agent's effort level is the same for the case in which the observed precursor cannot be concealed and for the case in which it can be concealed. In all other cases, the effort level is higher in the case in which concealment is possible.

In Case 1, the agent believes that project execution is acceptable to the point where the accident probability is higher than that of the DM ($x_1^* < x_1^A$). The DM stops the project if no precursor is shown. If a precursor $x_1 \in [0, x_1^*]$ is obtained, the agent indicates the precursor and encourages the DM to proceed with the project. However, if a precursor is observed such that $x_1 \in (x_1^*, x_1^A]$, concealing the observed value cannot change the DM's decision to align with the agent's interests. Meanwhile, if any precursor is not obtained, the project is halted. Therefore, the agent tries to obtain an observed value as much as possible by exerting effort to carry out the project.

In Case 2, the agent believes that the project should be allowed to proceed only to the point at which the accident probability is lower than that of the DM ($x_1^* > x_1^A$). The DM carries out the project if no precursor is shown. If a precursor is obtained such that $x_1 \in [0, x_1^A]$, the agent indicates the precursor and encourages the DM to proceed with the project. However, if the observed value is obtained such that $x_1 \in (x_1^A, x_1^*)$, the DM's decision cannot be changed by concealment, as in Case 1. Meanwhile, if any precursor is not obtained, the project is carried out. This is no different from the case in which secrecy is not possible. Therefore, the incentive for effort is the same as the case in which concealment is not possible.

In Case 3, the agent believes that project execution is acceptable to the point at which the accident probability is higher than that of the DM ($x_1^* < x_1^A$). The DM carries out the project if no precursor is shown. If the observed precursor is such that $x_1 \in [0, x_1^*]$, the agent shows it to the DM to encourage the execution of the project, while if $x_1 \in (x_1^*, x_1^A)$, the project is stopped if the value is shown to the DM; however, by keeping the information concealed, the DM's decision can be changed to execute a project that the agent considers desirable. This increases the value of the project to the agent. Meanwhile, if any precursor is not obtained, the project is carried out, and this is no different from the case in which the information cannot be kept secret. Thus, the incentive for effort is greater than in the case in which concealment is not possible.

In Case 4, the agent believes that the project should be allowed to proceed only to the point at which the accident probability is lower than that of the DM ($x_1^* > x_1^A$). The DM stops the project if no precursor is shown. If the observed precursor is such that $x_1 \in [0, x_1^A]$, the agent shows it to the DM to encourage the project to proceed, while if $x_1 \in (x_1^A, x_1^*]$, the project proceeds if the observed value is shown to the DM, but by withholding the information, the DM's decision can be changed to halt the project, as the agent deems desirable. This increases the value of the project to the agent. Meanwhile, the project is not executed if the precursor is not obtained, and this is no different from the case in which the information cannot be kept secret. Thus, the incentive for effort is greater than in the case in which concealment is not possible.

Thus, when the DM decides to halt the project if the information is not available, the agent intends to avoid this situation and increases the effort level to obtain an observed value. Furthermore, if the agent believes that the DM's decision can be changed by concealment, the project value increases by the ability to conceal information. Therefore, the incentive to realize the project also increases, and the level of effort also increases. Increased value of the information stimulates the incentive to obtain it. Meanwhile, the incentive to obtain the precursor is reduced by the possibility of concealment only when the DM decides to halt the project if the precursor is not shown. The incentive is reduced because, with no information, the agent believes that the project is too risky to proceed and halts the project by neglecting to make the necessary effort.

6.4 Case of Optimistic Agent

In this section, I explain that Cases 1 and 3 may be equilibria, along with the nature of each equilibrium. Cases 1 and 3 have the following relationship in common:

$$D^A \leq \frac{B+R}{B} \cdot D^M$$

in the accidental damage beliefs of the DM and the agent. In other words, the accident damage belief per benefit of the project, including private benefits, is lower for the agent than for the DM. This is the case when the agents are optimistic in that sense. When these equilibria exist, the effort level is higher in the Case 1 equilibrium than in the Case 3 equilibrium, as shown in the following Lemma 6.5.

Lemma 6.5 (Effort levels in Cases 1 and 3)

$$e_1 > e_3 > 0 \text{ when } D^A \in \left(B+R, \frac{B+R}{B} D^M\right]$$

$$e_1 > e_3 = 0 \textit{ when } D^A \in (0, B+R]$$

Proposition 6.2 demonstrates the existence of values for certain parameters such that each case is an equilibrium.

Proposition 6.2 (Separating and pooling equilibrium for optimistic agent) *Under Assumptions 3.1, 3.2, 3.3, 3.4, 4.1, and 4.2, Case 1 is an equilibrium when* $\tau \to \left(1 - B/D^M\right)^+$, *and Case 3 is an equilibrium when* $\tau \to 1^-$.

In Cases 1 and 3, the agent believes that the project should be allowed to proceed to the point at which the probability of an accident is higher than that of the DM ($x_1^* < x_1^A$). In Case 1, the DM stops the project if no precursor is shown, and thus, concealment cannot change the DM's decision. The equilibrium in this case is a separating equilibrium. In this model, the agent is not explicitly compensated other than a fixed private profit. Therefore, it is not possible to optimize the compensation structure to form a separating equilibrium as in the usual agency model. Only the DM's decision-making corresponds to compensation. By setting the DM's decision as "stopping the project when there is no information," the concealment strategy of the agent is invalidated.

In Case 3, meanwhile, the DM carries out the project when no precursor is shown. The DM cannot identify whether the precursor is not observed or whether it is concealed. Nevertheless, by deciding to proceed with the project, the DM can expect a higher project value. This equilibrium is a pooling equilibrium. In fact, in the case $\tau \to 1^-$, the probability of an accident occurring is very small, but if the precursor is not obtained, choosing Case 1 halts the project with a certain probability. If this loss exceeds the loss from the agent's concealment of the observed precursor, Case 3 is chosen.

In each equilibrium, the agent's effort levels e_i $(i = 1, 3)$ are affected by the agent's accident damage estimate D^A and the size of its private profit R, as the following Proposition 6.3 shows.

Proposition 6.3 (Effort incentive for optimistic agent)

$$\frac{de_1}{dD^A} < 0, \ \frac{de_1}{dR} > 0, \ \frac{de_3}{dD^A} > 0, \ \frac{de_3}{dR} < 0$$

The last two inequalities hold when $e_3 > 0$.

Thus, in the separating equilibrium (Case 1), more optimistic agents work harder, while private profit stimulates the incentive to work harder. Meanwhile, in the pooling equilibrium (Case 3), more optimistic agents are less willing to work harder, and the existence of private profit reduces the incentive to work harder.

In the separating equilibrium, the agent strives to increase the probability of obtaining the precursor because the project is stopped if any precursor is not shown. The agent with smaller accident damage estimate works harder because he believes that the project value will be higher if it is carried out. The same effect is expected when the private profit is larger, as it increases the project value for the agent as well.

In the pooling equilibrium, project execution is ensured even when no precursor is obtained. As the optimistic agent's expected accident damage is smaller, the difference between the project value with no observed information and that with observed information becomes smaller, resulting in lower information value as information accuracy increases. Meanwhile, the greater the private profit, the more distorted the search for more accurate information becomes. This means that the presence of private profit hinders the incentive to put in effort.

6.5 Case of Pessimistic Agent

In this section, I explain that Cases 2 and 4 may be equilibria, along with the nature of each equilibrium. Cases 2 and 4 have the following relationship in common:

$$D^A > \frac{B + R}{B} \cdot D^M.$$

in the accidental damage beliefs of the DM and the agent. In other words, the accident damage belief per benefit of the project, including private benefits, is higher for the agent than for the DM. This is the case when the agent is pessimistic in that sense.

When these equilibria exist, the following Lemma 6.6 shows the effort level of the agent.

Lemma 6.6 (Effort levels in Cases 2 and 4)

$$e_4 > e_2 > 0 \text{ when } D^A \to \left(\frac{B+R}{B}D^M\right)^+, e_4 < e_2 > 0 \text{ when } D^A \to \infty,$$

$$e_4 > e_2 > 0 \text{ when } D^A > \frac{B+R}{B}D^M \text{ and } W_0^A > 0.$$

The following Proposition 6.4 shows that values for parameters exist for which Cases 2 and 4 are equilibria.

Proposition 6.4 (Separating and pooling equilibria for pessimistic agent) *Under Assumptions 3.1, 3.2, 3.3, 3.4, 4.1, and 4.2, Case 2 is an equilibrium when* $\tau \to 1^-$, *and Case 4 is an equilibrium when* $\tau \to \left(1 - \frac{B}{D^M}\right)^+$ *and* $D^A \to \frac{B+R}{B}D^M$.

In Cases 2 and 4, the agent believes that the project should only be allowed to proceed to the point at which the accident probability is lower than that of the DM ($x_1^* > x_1^A$). In Case 2, the DM executes the project if the precursor is not shown, and thus, concealment cannot change the DM's decision when a precursor is obtained. Therefore, the equilibrium in this case is a separating equilibrium. As in Case 1, by setting the DM's decision as "carrying out the project in the absence of information," the agent's strategy of secrecy is invalidated.

Meanwhile, in Case 4, the DM stops the project if no precursor is shown. The DM cannot discern whether the precursor is not obtained or whether it is kept secret. Nevertheless, the DM can still expect a higher project value by deciding to halt the project. This equilibrium is a pooling equilibrium. In the case $\tau \to 1 - B/D^M$, the probability of an accident is large and the expected value of the damage is close to the project benefit. If Case 4 is chosen when no precursor is obtained, the agent is likely to conceal an observed value that is unacceptable for the DM. Case 4 is an equilibrium when this loss does not exceed the project benefit.

In each equilibrium, the agent's effort levels e_i $(i = 2, 4)$ are affected by the agent's expected accident damage D^A and the size of its private profit R, as the following Proposition 6.5 shows.

Proposition 6.5

$$\frac{de_2}{dD^A} > 0, \ \frac{de_2}{dR} < 0, \ \frac{de_4}{dD^A} < 0, \ \frac{de_4}{dR} > 0$$

In the separating equilibrium (Case 2), more pessimistic agents work harder, while private benefits reduce their incentive to work harder. In the pooling equilibrium (Case 4), more pessimistic agents work less, while the existence of private profit stimulates the incentive to work harder.

In the separating equilibrium, the project is carried out even if no precursor is shown. For a more pessimistic agent, the expected damage from an accident is higher. To avoid this, the more pessimistic agent tries to increase the probability of obtaining a precursor. However, when the private profit is large, the agent's private profit compensates for a part of the damage and then works to inhibit his effort. In the pooling equilibrium, the project is halted if the precursor is not shown. A more pessimistic agent with higher accident damage expectation seeks to avoid the possibility that the precursor is obtained and the project is carried out. They tend to avoid exerting effort to do so. Meanwhile, the private profit increases the project value for the agent when the project is carried out, thereby mitigating the above effect.

6.6 Conclusion

What implications does the analysis in this chapter have for nuclear policy? This chapter analyzes models that consider small accidents in the operation of nuclear power as precursors to major accidents. It also explores whether the possibility that utilities could cover up such signals of danger could hinder utilities' risk assessment. The answer to this question is negative. That is, the ability to cover up is not in itself a reason for not actively conducting risk assessments. In the course of this explanation, if the occurrence of a precursor minor accident could be covered up, there are four possible states (Cases 1 to 4). Assuming the power utility has a lower, more optimistic estimate of damages in the event of an accident than the regulator, then this is limited to Cases 1 and 3. At this point, as the regulator, two policy options exist for the utility when it is not presented with sufficient evidence of safety: cease generation or enforce it. In the absence of sufficient evidence, Case 1 is to suspend nuclear power generation and Case 3 is to enforce it. If, as we saw in Chapter 1, nuclear power was promoted without sufficient safety evidence, Case 3 would seem to be the closest to reality in Japan.

The perception that Case 3 is closest to reality is tantamount to the regulator choosing a pooled equilibrium, as indicated in this chapter. If the regulator chooses Case 1 by adopting a policy of discontinuing nuclear power generation unless sufficient safety evidence is presented, the possibility arises that nuclear power generation will be discontinued even though it is worth continuing. When the losses from this type of error are expected to be significant for society, Case 3 is chosen. In Case 3, the ability to cover up increases the value of the nuclear project to the utility relative to the case where it cannot be covered up, and the incentive to obtain evidence can be expected to increase. However, the value of information obtained relative to that when it is not obtained is not large, and incentives are further reduced under large private benefits. If Case 1 had been chosen and the policy had been to shut down nuclear power generation unless sufficient safety evidence had been accumulated, the more optimistic operators would have made a greater effort for risk assessment to present concrete evidence. The operators with greater private interests would have pursued research and development to obtain more favorable evidence for the continuation of nuclear power.

Chapter 7
Conclusion of Part 1

If the decision to promote the use of nuclear energy is made without sufficient risk assessment, what factors could have contributed to this inadequate assessment? To address this issue, I develop a model in which the agent managing the project incurs a cost to gather information for risk assessment; the DM evaluates the social costs and benefits of the project based on this information, and decides whether to execute the project. Although the model used in the analysis is a simple one, many factors are analyzed. The main factors identified in the analysis that deter the effort for risk assessment include optimism about the risk of the operating agent of the project, large private profit associated with the business, discrepancy between the expected magnitude of the risk in the agent and the criteria required by society, large difference in risk expectations between the agent and the DM, and ambiguous liability. Meanwhile, factors that do not deter the operating agent from effort in risk assessment include: the ability to capture the DM and make decisions independently of the DM, and the ability to keep risk-related information confidential.

In the model, the DM, which lacks expertise, outsources the gathering of information to experts. There have been several previous models that examine communication issues within this framework, but to the best of my knowledge, none of them have analyzed the problem of risk assessment. As a result, the analysis of this study is distinct from the previous models. However, some of the propositions examined in previous communication models are confirmed. For example, I confirm that non-biased agents are more likely to expend effort even when the acquired information is publicly verifiable.

There are several limitations to the scope of this study. First, when the technology is not yet perfected and the project relies on trial and error, it is necessary to conduct a sequential risk assessment and decision-making for the project to progress. Second, a different problem arises when the project execution and risk assessment must be conducted simultaneously, as it is much more difficult to stop a project in the middle than to decide on execution after assessing risks in advance. This issue will be addressed in the second part of this book. Additionally, this model is not

A. Torii, *Decision-making in High-Risk Projects*,
https://doi.org/10.1007/978-981-97-3056-8_7

appropriate for analyzing optimal decision-making or project execution, as it is focused on understanding why information-gathering efforts were insufficient in nuclear development and identifying the factors that contributed to this inadequacy. In this regard, the approaches discussed in Chap. 2, which analyze cheap-talk models involving multiple experts and the role of advocates, may be instructive. For example, it may be possible to gain insights into the decision-making process when there are multiple agency companies executing a project by analyzing these models.

Appendix: Proofs for Propositions and Lemmata

Proof of Lemma 3.1 Under continuous uniform distribution with parameter θ, the probability that all values of n random variables X falls in the region less than y^* is $(y^*/\theta)^n$, when $0 \leq y^* \leq \theta$. This probability represents the distribution function of the random variable, y_n. Then, denoting the probability density function of y_n as $g(y|\theta)$,

$$g(y_n|\theta) = \begin{cases} \dfrac{d}{dy_n}\left(\dfrac{y_n}{\theta}\right)^n = \dfrac{n\,(y_n)^{n-1}}{\theta^n} & \text{for } 0 \leq y \leq \theta \\ 0 & \text{otherwise.} \end{cases}$$

By Bayes' theorem,

$$f(\theta|\boldsymbol{x}_1) = \frac{g(y_1|\theta)f(\theta|\boldsymbol{x}_0)}{\displaystyle\int_0^1 g(y_1|\theta')f(\theta'|\boldsymbol{x}_0)d\theta'} = \begin{cases} -\dfrac{1}{\theta\log(x_1)} & \text{for } x_1 \leq \theta \leq 1 \\ 0 & \text{otherwise} \end{cases}$$

for case $n = 1$, and

$$f(\theta|\boldsymbol{x}_2) = \frac{g(y_2|\theta)f(\theta|\boldsymbol{x}_0)}{\displaystyle\int_0^1 g(y_2|\theta')f(\theta'|\boldsymbol{x}_0)d\theta'} = \begin{cases} -\dfrac{1}{\theta^2\left((y_2)^{-1}-1\right)} & \text{for } y_2 \leq \theta \leq 1 \\ 0 & \text{otherwise} \end{cases}$$

for case $n = 2$. Integrating this probability density function gives the distribution function. From this posterior probability distribution, the subjective probability of an accident occurring is

$$p(\theta > \tau|\boldsymbol{x}_1) = 1 - F(\tau|\boldsymbol{x}_1) = \begin{cases} \dfrac{\log(\tau)}{\log(x_1)} & \text{for } 0 \leq x_1 \leq \tau \\ 1 & \text{for } \tau < x_1 \leq 1 \end{cases}$$

for case $n = 1$, and

$$p(\theta > \tau|\boldsymbol{x}_2) = 1 - F(\tau|\boldsymbol{x}_2) = \begin{cases} \dfrac{\left(\tau^{-1} - 1\right)}{\left((y_2)^{-1} - 1\right)} & \text{for } 0 \leq y_2 \leq \tau \\ 1 & \text{for } \tau < y_2 \leq 1 \end{cases}$$

for case $n = 2$. □

Proof of Lemma 3.2 The condition σ_n^* should maximize the function (3.2) pointwisely:

$$\sigma_n^* = \{\boldsymbol{x}_n | B - p(\theta > \tau|\boldsymbol{x}_1) \cdot D^M \geq 0\}.$$

Here, condition $B - p(\theta > \tau|\boldsymbol{x}_1) \cdot D_M \geq 0$ is transformed to condition $y_n \leq x_n^*$ by Lemma 3.1, $n = \{1, 2\}$, where x_n^* is defined as the solution of the equation $h_n(x) = B/D^M$ and

$$h_n(x) \equiv \begin{cases} \dfrac{\log(\tau)}{\log(x)} & \text{for } n = 1 \\ \dfrac{\tau^{-1} - 1}{x^{-1} - 1} & \text{for } n = 2. \end{cases}$$

For $x \in [0, \tau]$,

$$h_1'(x) = -\frac{\log(\tau)}{x\,(\log(x))^2} > 0, \quad \lim_{x \to 0} h_1(x) < 0 < B/D^M, \quad h_1(\tau) = 1 > B/D^M$$

$$h_2'(x) = \frac{\tau^{-1} - 1}{x^2\left(x^{-1} - 1\right)^2} > 0, \quad \lim_{x \to 0} h_2(x) = 0 < B/D^M, \quad h_2(\tau) = 1 > B/D^M,$$

then $x_n^* < \tau$. Define a function:

$$h_3(\tau, x) \equiv \frac{\log(\tau)}{\log(x)} - \frac{\tau^{-1} - 1}{x^{-1} - 1}.$$

As

$$\frac{\partial}{\partial \tau} h_3(\tau, x) = -\frac{x \log(x) + \tau(1 - x)}{\tau^2(x - 1)\log(\tau)},$$

$$\frac{\partial}{\partial \tau} h_3(\tau, x) \begin{cases} > 0, & \text{for } x < \tau < \tau^* < 1, \\ < 0, & \text{for } x < \tau^* < \tau < 1, \end{cases}$$

where

$$\tau^* \equiv -\frac{x \log(x)}{1 - x}, \quad \tau^* \in (x, 1),$$

as

$$\tau^* - x = -\frac{x\,(1 - x + \log(x))}{1 - x} > 0, \tau^* - 1 = -\frac{1 - x + x\log(x)}{1 - x} < 0.$$

Therefore,

$$h_3(\tau, x) \geq \min\left(h_3(x, x), h_3(1, x)\right) = \min(0, 0) = 0.$$

Thus, given $\tau < 1$ the locus of the function $h_1(x)$ locates above the locus of the function $h_2(x)$ for the region $x \in (0, \tau)$. Then, the solution of the equation $h_1(x) = B/D^M$ is less than the solution of the equation $h_2(x) = B/D^M$.

Furthermore,

$$\frac{\partial x_1^*}{\partial D^M} = \tau^{D^M/B}\frac{\log(\tau)}{B} < 0, \ \frac{\partial x_2^*}{\partial D^M} = -\tau\delta^2\frac{1-\tau}{B} < 0.$$

□

Proof of Lemma 3.3

$$\gamma_n = \frac{p\left((y_n > x_n^*) \wedge (\theta \leq \tau)|\boldsymbol{x}_0\right)}{p(\theta \leq \tau|\boldsymbol{x}_0)} = \frac{1}{\tau}\int_{x_n^*}^{\tau}\left(1 - \left(\frac{x_n^*}{\theta}\right)^n\right)d\theta$$

$$\beta_n = \frac{p\left((y_n \leq x_n^*) \wedge (\theta > \tau)|\boldsymbol{x}_0\right)}{p(\theta > \tau|\boldsymbol{x}_0)} = \frac{1}{1-\tau}\int_{\tau}^{1}\left(\frac{x_n^*}{\theta}\right)^n d\theta$$

for $n = 1,\ 2$. □

Proof of Remark 3.1 Omitted for brevity. □

Proof of Lemma 4.1 $\sigma_1^* = \{x_1|x_1 \in [0, x_1^*]\}$ by Lemma 3.2, and $p(\theta > \tau|x_1)$ by Lemma 3.1 gives the value of $W_1^a(x_1)$. From the subjective probability based on $F(\theta|\boldsymbol{x}_0)$,

$$\begin{aligned}
E_{x_1}(W_1^a(x_1)|\boldsymbol{x}_0) &= \int_0^1\int_0^{\min(\theta, x_1^*)}\left(B + R - D^A\frac{\log(\tau)}{\log(x_1)}\right)\frac{dx_1}{\theta}dF(\theta|\boldsymbol{x}_0)\\
&= \int_0^{x_1^*}\int_{x_1}^{1}\left(B + R - D_A\frac{\log(\tau)}{\log(x_1)}\right)\frac{d\theta}{\theta}dx_1\\
&= \int_0^{x_1^*}\left(-(B+R)\log(x_1) + D^A\log(\tau)\right)dx_1\\
&= -(B+R)(x_1^*(\log(x_1^*) - 1)) + D^A x_1^*\log(\tau).
\end{aligned}$$

□

Proof of Lemma 4.2 Define $k^0(x)$ and $k^1(x)$ as $k^0(x) \equiv 1 - x + x\log(x)$ and $k^1(x) \equiv 1 - x + \log(x)$, then $k^0(1) = k^1(1) = 0$, and $k^{0'}(x) = \log(x) < 0$, $k^{1'}(x) = 1/x - 1 > 0,\ \ \forall x \in (0, 1)$. So

$$k^0(x) = 1 - x + x\log(x) > 0,\ \ k^1(x) = 1 - x + \log(x) < 0,\ \ \forall x \in (0, 1). \quad \text{(A.1)}$$

$$W_1^A - W_0^A = D^A\left\{1 - \tau + x_1^*\log(\tau)\right\} - (B + R)\left\{1 - x_1^* + x_1^*\log(x_1^*)\right\}.$$

As $x_1^* = \tau^{D^M/B} < \tau$ and $\tau \in (0, 1)$, $1 - \tau + x_1^*\log(\tau) > 1 - \tau + \tau\log(\tau) > 0$ ($\because$ $(A.1)$). Then, $W_1^A \gtreqless W_0^A \iff D^A \gtreqless \bar{D}^A$.

Define $k^2(x_1^*)$ and $k^3(x_1^*)$ as

$$\begin{aligned}
k^2(x_1^*) &\equiv \frac{\log(\tau)}{\log(x_1^*)}\frac{1 - x_1^* + x_1^*\log(x_1^*)}{1 - \tau + x_1^*\log(\tau)} \\
k^3(x_1^*) &\equiv x_1^*(1-\tau)\left(\log(x_1^*)\right)^2 - x_1^*\log(x_1^*)\left(1 - \tau + \log(\tau)\right) \\
&\quad - x_1^*(1 - x_1^*)\log(\tau) - (1-\tau)(1 - x_1^*).
\end{aligned}$$

Then,

$$\begin{aligned}
\bar{D}^A &= D^M\frac{B+R}{B}k^2(x_1^*), \\
\frac{dk^2(x_1^*)}{dx_1^*} &= \frac{\log(\tau)k^3(x_1^*)}{x_1^*\left(\log(x_1^*)\right)^2\left(1 - \tau + x_1^*\log(\tau)\right)^2}, \\
\frac{dk^3(x_1^*)}{dx_1^*} &= (1-\tau)\left(\log(x_1^*)\right)^2 + (1 - \tau - \log(\tau))\log(x_1^*) - 2(1 - x_1^*)\log(\tau), \\
\frac{d^2k^3(x_1^*)}{dx_1^{*2}} &= \frac{1}{x_1^*}\left\{2(1-\tau)(1-x_1^*)\left(\frac{\log(x_1^*)}{1 - x_1^*} - \frac{\log(\tau)}{1-\tau}\right)\right. \\
&\quad \left. +1 - \tau + \log(\tau)\right\} < 0.
\end{aligned}$$

Note that function $\log(x)/(1-x)$ is an increasing function for $x \in (0, 1)$ and by Eq. (A.1). Here,

$$\begin{aligned}
\left.\frac{dk^3(x_1^*)}{dx_1^*}\right|_{x_1^*=\tau} &= -\log(\tau)\left(1 - \tau + \tau\log(\tau)\right) > 0, \\
k^3(\tau) &= -\left(1 - \tau + \tau\log(\tau)\right)^2 < 0.
\end{aligned}$$

Therefore, $k^3(x_1^*) < 0$, then, $k^{2'}(x_1^*) > 0$, $\forall x_1^* \in (0, \tau)$. By Assumption 3.4, $D^M/B < 1/(1-\tau)$, and then, $x_1^* = \tau^{D^M/B} > \tau^{\frac{1}{1-\tau}}$. Then,

$$D^M\frac{B+R}{B}k^2\left(\tau^{\frac{1}{1-\tau}}\right) < \bar{D}^A < D^M\frac{B+R}{B}k^2(\tau) = D^M\frac{B+R}{B}.$$

Define function

$$k^4(\tau) \equiv \frac{-(1-\tau-\log(\tau)) + \tau^{-\frac{1}{1-\tau}} - \tau^{-\frac{\tau}{1-\tau}}}{\log(\tau) + \tau^{-\frac{1}{1-\tau}} - \tau^{-\frac{\tau}{1-\tau}}}.$$

Then, $k^2\left(\tau^{\frac{1}{1-\tau}}\right) = k^4(\tau)$, and

$$\frac{dk^4(\tau)}{d\tau} = -\frac{(1-\tau+\tau\log(\tau))\left(\tau^{-\frac{1}{1-\tau}} - 1\right)}{\tau\left(\log(\tau) + \tau^{-\frac{1}{1-\tau}} - \tau^{-\frac{\tau}{1-\tau}}\right)^2} < 0.$$

Thus, $k^4(\tau) > k^4(1)$, and $\lim_{\tau\to 1} k^4(\tau) = (\exp(1)-2)/(\exp(1)-1)$. □

Proof of Proposition 4.1 By assumption, $D^A > \bar{D}^A$, then $W_1^A > W_0^A$ from Lemma 4.2. Define $l(e) \equiv c'(e)/\pi'(e)$, then $l(0) = 0$, $\lim_{e\to\bar{e}} l(e) \to \infty$, and $l'(e) = (c''(e)\pi'(e) - c'(e)\pi''(e))/(\pi'(e))^2 > 0$ on $e \in [0, \bar{e}]$. Therefore, equation $l(e) = W_1^A - W_0^A$ for e has a unique solution $e^* \in [0, \bar{e})$. Define the following function:

$$\begin{aligned} k^5(R, D^M, D^A) &\equiv W_1^A - W_0^A \\ &= D^A\left(1-\tau+x_1^*\log(\tau)\right) - (B+R)\left(1-x_1^*+x_1^*\log(x_1^*)\right). \end{aligned}$$

Then, differentiating the condition (4.3) gives

$$\frac{\partial e^*}{\partial D^A} = -\frac{\pi'(e^*)\cdot\dfrac{\partial k^5(R, D^M, D^A)}{\partial D^A}}{\pi''(e^*)\cdot k^5(R, D^M, D^A) - c''(e^*)} > 0.$$

Note that $\partial k^5/\partial D^A = 1-\tau+x_1^*\log(\tau) > 0$, as shown in the proof of Lemma 4.2. In the same manner,

$$\frac{\partial}{\partial R}k^5(R, D^M, D^A) = -\left(1-x_1^*+x_1^*\log(x_1^*)\right) < 0, \quad \forall x_1^* \in (0,1)$$

leads $\partial e^*/\partial R < 0$. □

Proof of Proposition 4.2 First, $D^M > B\bar{D}^A/(B+R)$ is demonstrated as

$$\frac{D^M}{B} = \frac{\log(x_1^*)}{\log(\tau)} > \frac{1-x_1^*+x_1^*\log(x_1^*)}{1-\tau+x_1^*\log(\tau)} = \frac{\bar{D}^A}{B+R}.$$

Here,

$$\frac{\log(x_1^*)}{\log(\tau)} - \frac{1 - x_1^* + x_1^* \log(x_1^*)}{1 - \tau + x_1^* \log(\tau)} = \frac{(1-\tau)(1-x_1^*)}{\log(\tau)\left(1 - \tau + x_1^* \log(\tau)\right)} \left(\frac{\log(x_1^*)}{1 - x_1^*} - \frac{\log(\tau)}{1-\tau} \right) > 0.$$

Then, at $D^M = BD^A/(B+R)$, $D^A = (B+R) \cdot D^M/B > \bar{D}^A$. By Lemma 4.2, this means $W_1^A > W_0^A$, and by Proposition 4.1, $e^* > 0$.

For given D^A, if D^M is such that $D^A \leq \bar{D}^A$, $e^* = 0$ as $W_0^A \geq W_1^A$ by Lemma 4.2, while if $D^A > \bar{D}^A$, $e^* > 0$ and $W_0^A < W_1^A$. In the latter case, differentiating the condition (4.3) gives

$$\frac{\partial e^*}{\partial D^M} = -\frac{\pi'(e^*) \cdot \dfrac{\partial k^5(R, D^M, D^A)}{\partial D^M}}{\pi''(e^*) \cdot k^5(R, D^M, D^A) - c''(e^*)},$$

where the function $k^5(\cdot)$ is that defined in the proof of Proposition 4.1. As

$$\frac{\partial k^5}{\partial D^M} = \frac{\partial k^5}{\partial x_1^*} \frac{dx_1^*}{dD^M} = \frac{x_1^*}{B} (\log(\tau))^2 \left(D^A - \frac{B+R}{B} D^M \right),$$

$e^* > 0$ is maximized at $D^M = BD^A/(B+R)$, and as shown above, the combination $(D^M, D^A) = (BD^A/(B+R), D^A)$ falls in the region $e^* > 0$. □

Proof of Proposition 4.3 Given value of $D^M = D^A$ falls in the region $D^M = D^A > \bar{D}^A$, as $e^* > 0$. The function $k^2(x_1^*)$ defined in the proof of Lemma 4.2 represents the ratio $\bar{D}^A/D^M \cdot B/(B+R)$ and shows the increasing function of x_1^*, and then the decreasing function of D^M. Decreasing this ratio implies that inequality $D^M = D^A > \bar{D}^A$ holds for the increasing value of $D^M = D^A$. For the function $k^5(\cdot)$ in the proof of Proposition 4.1

$$\left. \frac{\partial k^5}{\partial D^A} + \frac{\partial k^5}{\partial D^M} \right|_{D^M = D^A} = 1 - \tau + x_1^* \log(\tau) - x_1^* \log(\tau) \log(x_1^*) \frac{R}{B},$$

which has a positive value when $R/B < (1 - \tau + x_1^* \log(\tau))/(x_1^* \log(\tau) \log(x_1^*))$. □

Proof of Lemma 5.1 From Lemma 3.2, the project is executed when $y_2 \leq x_2^*$ and halted when $y_2 > x_2^*$. Depending on the value of the parameter θ, the probability of the project being conducted is 1 when $\theta \leq x_2^*$, and x_2^*/θ when $\theta > x_2^*$. Therefore,

$$E_{\boldsymbol{x}_2}(H_{\sigma_2^*}(\boldsymbol{x}_2)|\boldsymbol{x}_1) = \int_{x_1}^{x_2^*} dF(\theta|\boldsymbol{x}_1) + \int_{x_2^*}^{1} \frac{x_2^*}{\theta} dF(\theta|\boldsymbol{x}_1)$$
$$= -\frac{1}{\log(x_1)}\left(\log\left(\frac{x_2^*}{x_1}\right) + 1 - x_2^*\right).$$

On the contrary, since the probability that the project will be executed and the accident will occur is 1 when $\theta > \tau$,

$$E_{\boldsymbol{x}_2}(H_{\sigma_2^*}(\boldsymbol{x}_2) \cdot p(\theta > \tau|\boldsymbol{x}_2)|\boldsymbol{x}_1) = \int_{\tau}^{1} \frac{x_2^*}{\theta} dF(\theta|\boldsymbol{x}_1) = -\frac{x_2^*}{\log(x_1)} \cdot \frac{1-\tau}{\tau}$$

holds. Therefore, the project value is

$$E_{\boldsymbol{x}_2}(H_{\sigma_2^*}(\boldsymbol{x}_2) \cdot (B + R - p(\theta > \tau|\boldsymbol{x}_2) \cdot D_A)\,|\boldsymbol{x}_1)$$
$$= (B+R)E_{\boldsymbol{x}_2}(H_{\sigma_2^*}(\boldsymbol{x}_2)|\boldsymbol{x}_1) - D_A E_{\boldsymbol{x}_2}(H_{\sigma_2^*}(\boldsymbol{x}_2) \cdot p(\theta > \tau|\boldsymbol{x}_2)|\boldsymbol{x}_1)$$
$$= -\frac{1}{\log(x_1)}\left\{(B+R)\cdot\left(\log\left(\frac{x_2^*}{x_1}\right) + 1 - x_2^*\right) - D_A \cdot x_2^* \cdot \frac{1-\tau}{\tau}\right\}.$$

The relations $x_2^* = \delta\tau$ from Lemma 3.2 and $(1-\delta\tau)B = \delta(1-\tau)D^M$ lead the Lemma. □

Proof of Lemma 5.2 Define a function

$$\mu_1(D^M, D^A; R, \tau) \equiv (B+R)(1 - \delta\tau + \log(\delta\tau)) - D^A(\delta(1-\tau) + \log(\tau)).$$

When $x_1 \in (0, \min(x_1^*, x_2^*) = x_1^*)$,

$$W_2^a(x_1) - W_1^a(x_1) = \frac{\mu_1(D^M, D^A; R, \tau)}{-\log(x_1)}. \tag{A.2}$$

The function $\mu_1(D^M, D^A; R, \tau)$ is defined as a function of D^M and D^A for the analysis and abbreviated as $\mu_1(D^M, D^A)$. Define another function

$$\mu_2(D; R, \tau) \equiv \mu_1(D, D; R, \tau)$$

that is abbreviated as $\mu_2(D)$. For the function $\mu_2(\cdot)$,

$$\mu_2(B) = R(1 - \tau + \log(\tau)) < 0 \ \ (\because\ \tau \in (0, 1)).$$

Note that when $D^M = B, \delta = 1$. As

$$D^M = \frac{B}{1-\tau} \Leftrightarrow \delta = \frac{1}{1+\tau},\ \ \frac{1}{1+\tau} + \log\left(\frac{\tau}{1+\tau}\right) < 0,\ \text{for } \tau \in (0, 1),$$

we obtain

$$\begin{aligned}\mu_2\left(\frac{B}{1-\tau}\right) &= (B+R)\left(\frac{1}{1+\tau}+\log\left(\frac{\tau}{1+\tau}\right)\right)-\frac{B}{1-\tau}\left(\frac{1-\tau}{1+\tau}+\log(\tau)\right)\\ &= R\left(\frac{1}{1+\tau}+\log\left(\frac{\tau}{1+\tau}\right)\right)+B\left(\log\left(\frac{\tau}{1+\tau}\right)-\frac{\log(\tau)}{1-\tau}\right)\\ &> B\left(\frac{1}{1+\tau}+\log\left(\frac{\tau}{1+\tau}\right)\right)+B\left(\log\left(\frac{\tau}{1+\tau}\right)-\frac{\log(\tau)}{1-\tau}\right)\\ &= B\left(\frac{1}{1+\tau}+2\log\left(\frac{\tau}{1+\tau}\right)-\frac{\log(\tau)}{1-\tau}\right). \qquad \text{(A.3)}\end{aligned}$$

Inside the parenthesis in the Eq. (A.3) takes a positive value for $\tau > \tau^+$. As the function $\mu_2(\cdot)$ is continuous, when $\tau > \tau^+$, there exists $D_+^M \in (B, B/(1-\tau))$ such that for all $D \geq D_+^M$, $\mu_2(D) = \mu_1(D, D) \geq 0$.

Meanwhile, for the function $\mu_1(D^M, D^A)$

$$\begin{aligned}\frac{\partial \mu_1(D^M, D^A)}{\partial D^A} &= -\log(\tau) - \delta(1-\tau)\\ &> 1-\tau-\delta(1-\tau) = (1-\delta)(1-\tau) > 0,\\ \mu_1(D^M, B) &= B(1-\delta+\log(\delta)) + R(1-\delta\tau+\log(\delta\tau)) < 0.\end{aligned}$$

Therefore, when $\tau > \tau^+$, for $D^M > D_+^M$, there exists $D_+^A \in (B, D^M)$ such that if $D^A > D_+^A$, then $\mu_1(D^M, D^A) \geq 0$. Then, when $\tau > \tau^+$, $D^M > D_+^M$, and $D^A > D_+^A$, the numerator of the right-hand side of the Eq. (A.2) takes a positive value. Since the value of the function $\mu_1(D^M, D^A)$ is not dependent on the value of x_1, when $x_1 \to 0$ the right-hand side of the Eq. (A.2) converges to 0. As far as the numerator of Eq. (A.2) takes a positive value, evidently, the right-hand side is an increasing function of x_1. □

Proof of Lemma 5.3 As $R > 0$ and $x_2^* > x_1^*$ from Lemma 3.2, evidently $D_2^A > D_1^A > D^M$. When $D^A = D_1^A$, since $\tau > \tau^+$, $D^M > D_+^M$, $D^A = D_1^A > D^M > D_+^A(D^M)$,

$$\begin{aligned}W_2^a(x_1^*) &> W_1^a(x_1^*) = B + R - \frac{(B+R)D^M}{B}\frac{\log(\tau)}{\log(\tau^{D^M/B})} = 0,\\ W_2^a(x_2^*) &= B + R - (B+R)\frac{\log(\delta\tau)}{\log(\delta\tau)} = 0\end{aligned}$$

from Lemma 5.2. Meanwhile, for $x_1 \in (0, x_1^*]$,

$$\frac{\partial W_1^a(x_1)}{\partial D^A} = -\frac{\log(\tau)}{\log(x_1)} < 0,$$

and for $x_1 \in (0, x_2^*]$,

$$\frac{\partial W_2^a(x_1)}{\partial D^A} = \frac{\delta(1-\tau)}{\log(x_1)} < 0.$$

Therefore, for $x_1 \in (0, x_1^*]$,

$$\frac{\partial W_2^a(x_1)}{\partial D^A} - \frac{\partial W_1^a(x_1)}{\partial D^A} = \frac{\delta(1-\tau)+\log(\tau)}{\log(x_1)} > \frac{1-\tau+\log(\tau)}{\log(x_1)} > 0.$$

Moreover, when $D^A = D_2^A$,

$$W_2^a(x_1^*) = (B+R)\left(1 - \frac{1-\delta\tau+\log(\delta\tau)}{\log(x_1^*)}\right) + \frac{(B+R)D^M}{B}\left(1+\frac{\log(\delta\tau/x_1^*)}{1-\delta\tau}\right)\frac{\delta(1-\tau)}{\log(x_1^*)} = 0.$$

Then, when $D^A \geq D_1^A$, $W_1^a(x_1^*) \leq 0$ and $W_2^a(x_2^*) \leq 0$, when $D^A < D_1^A$, $W_1^a(x_1^*) > 0$ and $W_2^a(x_2^*) > 0$, when $D^A \geq D_2^A$, $W_2^a(x_1^*) \leq 0$, and when $D^A < D_2^A$, $W_2^a(x_1^*) > 0$. For $x_1 \in (0, x_2^*]$,

$$\frac{dW_2^a(x_1)}{dx_1} = \frac{1}{x_1(\log(x_1))^2}\big((B+R)(1-\delta\tau+\log(\delta\tau)) - D^A\delta(1-\tau)\big) < 0.$$

For the following use, let $\nu_1(x) \equiv W_2^a(x) - W_1^a(x) - \bar{c}$, $\nu_2(x) \equiv W_2^a(x) - \bar{c}$. The condition (5.1) is satisfied in the region where $\nu_1(x) \geq 0$ for $x \in (0, x_1^*]$, and in the region where $\nu_2(x) \geq 0$ for $x \in (x_1^*, x_2^*]$.

Let $\bar{c} < \min(W_2^a(x_1^*) - W_1^a(x_1^*), W_2^a(x_2^*))$. First, consider the case where $D^A \in (D_+^A(D^M), D_1^A)$. From Lemma 5.2, we have $D_+^A(D^M) < D^M$, and since $D^M < D_1^A$, this region exists. In this case, $W_2^a(x_1^*) - W_1^a(x_1^*) > 0$ and $W_2^a(x_2^*) > 0$. For $x_1 \in (0, x_1^*]$,

$$\nu_1'(x) > 0, \quad \lim_{x \to 0^+} \nu_1(x) = -\bar{c} < 0, \quad \nu_1(x_1^*) = W_2^a(x_1^*) - W_1^a(x_1^*) - \bar{c} > 0,$$

and for $x \in (0, x_2^*]$,

$$\nu_2'(x) < 0, \nu_2(x_1^*) = W_2^a(x_1^*) - \bar{c} > W_2^a(x_2^*) - \bar{c} = \nu_2(x_2^*) > 0.$$

Therefore, there exists $\chi_1 \in (0, x_1^*)$, such that if and only if $x_1 \in [\chi_1, x_1^*)$, $\nu_1(x_1) \geq 0$ and for all $x_1 \in [x_1^*, x_2^*]$, $\nu_2(x_1) \geq 0$.

Second, consider the case where $D^A \in [D_1^A, D_2^A)$. As $\bar{c} < W_2^a(x_1^*)$, for $x_1 \in (0, x_1^*]$,

$$\nu_1'(x) > 0, \quad \lim_{x \to 0^+} \nu_1(x) = -\bar{c} < 0,$$
$$\nu_1(x_1^*) = W_2^a(x_1^*) - W_1^a(x_1^*) - \bar{c} > W_2^a(x_1^*) - \bar{c} > 0,$$

and for $x \in (0, x_2^*]$,

$$\nu_2'(x) < 0, \nu_2(x_1^*) = W_2^a(x_1^*) - \bar{c} > 0, \ \nu_2(x_2^*) = W_2^a(x_2^*) - \bar{c} < -\bar{c} < 0.$$

Therefore, there exist $\chi_1 \in (0, x_1^*)$ and $\chi_2 \in (x_1^*, x_2^*)$, such that if and only if $x_1 \in [\chi_1, x_1^*)$, $\nu_1(x_1) \geq 0$ and if and only if $x_1 \in [x_1^*, \chi_2]$, $\nu_2(x_1) \geq 0$.

Lastly, consider the case where $D^A \in [D_2^A, \infty)$. As $\bar{c} < W_2^a(x_1^*) - W_1^a(x_1^*)$, for $x_1 \in (0, x_1^*]$,

$$\nu_1'(x) > 0, \quad \lim_{x \to 0^+} \nu_1(x) = -\bar{c} < 0, \ \nu_1(x_1^*) = W_2^a(x_1^*) - W_1^a(x_1^*) - \bar{c} > 0,$$

and for $x \in (0, x_2^*]$,

$$\nu_2'(x) < 0, \nu_2(x_1^*) = W_2^a(x_1^*) - \bar{c} < -\bar{c} < 0.$$

Therefore, there exists $\chi_1 \in (0, x_1^*]$ such that if and only if $x_1 \in [\chi_1, x_1^*]$, $\nu_1(x_1) \geq 0$ and for all $x_1 \in (x_1^*, x_2^*]$, $\nu_2(x_1) < 0$. □

Proof of Lemma 5.4 For a given observation value x_1 $(x_1 \in [0, 1])$, the parameter θ of the probability distribution that produces such an observation is limited to $\{\theta | \theta \in [x_1, 1]\}$. Therefore, the probability density of x_1 based on the no-information prior distribution is

$$\int_{x_1}^{1} p(x_1|\theta) dF(\theta|\varnothing) = \int_{x_1}^{1} \frac{1}{\theta} dF(\theta|\mathbf{x}_0) = -\log(x_1).$$

This lemma is obtained in conjunction with Lemma 5.3. □

Proof of Proposition 5.4 From Definition 5.5 and Eq. (A.2), χ_1 is the solution of the equation

$$W_2^a(\chi_1) - W_1^a(\chi_1) = -\frac{K}{\log(\chi_1)} = \bar{c} > 0,$$

where

$$K \equiv (B + R)(1 - \delta\tau + \log(\delta\tau)) - D^A(\delta(1 - \tau) + \log(\tau)).$$

As proven in Lemma 5.2, for $D^M > D_+^M$, $D^A > D_+^A(D^M)$, $\tau > \tau^+$, the K in the above equation takes a positive value. Therefore, as $\bar{c} \to 0^+$, $\chi_1 \to 0$. Moreover, differentiating the same equation with respect to D_A, we obtain

$$\frac{d\chi_1}{dD_A} = \frac{\chi_1}{\bar{c}}\{\delta(1-\tau)+\log(\tau)\} < \frac{\chi_1}{\bar{c}}\{\delta(1-\tau)+(\tau-1)\}$$
$$= \frac{\chi_1}{\bar{c}}\{(\delta-1)(1-\tau)\} < 0.$$

For $D^A \in (D^A_+(D^M), D^A_1)$ and $D^A \in [D^A_2, \infty)$, using the result of Lemma 5.4 and this derivative, when $\bar{c}$ is sufficiently small, we obtain

$$\frac{d}{dD^A}p(W_2^a(x_1) - W_1^a(x_1) \geq \bar{c}|\boldsymbol{x}_0) = \log(\chi_1)\frac{d\chi_1}{dD^A} > 0.$$

□

Proof of Lemma 5.5

$$Y_1^a(\bar{e}) = E_{x_1}\left(\max(W_1^a(x_1), W_1^a(x_1) + \lambda\left\{W_2^a(x_1) - W_1^a(x_1) - c\right\})|\boldsymbol{x}_0\right)$$
$$= W_1^A + \lambda E_{\boldsymbol{x}_0}\left(\max\left(0, W_2^a(x_1) - W_1^a(x_1) - c\right)|\boldsymbol{x}_0\right) \geq W_1^A$$

□

Proof of Proposition 5.5 First, assume $Y_1^a(\bar{e}) - \bar{c} \geq W_0^A$. This outcome is considered in two scenarios: when $W_1^A - \bar{c} \geq W_0^A$ and when $W_1^A - \bar{c} < W_0^A$. Let first consider the case where $W_1^A - \bar{c} \geq W_0^A$. If the agent does not make an effort in the first trial, he will try to obtain a precursor with effort for the second opportunity, resulting in the project value of $W_1^A - \bar{c}$. If the agent does exert effort in the first trial, the project value would be $Y_1^a(\bar{e})$. Therefore, the payoff for exerting effort is $W_1^A - \bar{c} + \lambda\left(Y_1^a(\bar{e}) - (W_1^A - \bar{c}) - \bar{c}\right)$. Given that $Y_1^a(\bar{e}) \geq W_1^A$ as per Lemma 5.5, this value is always at least $W_1^A - \bar{c}$. Consequently, the agent would opt to exert effort to obtain a precursor in the first trial opportunity.

Next, consider the case where $W_1^A - \bar{c} < W_0^A$. Similarly to the above, the decision to exert effort at the level of $\bar{e}$ in the first trial is based on a comparison between W_0^A and $W_0^A + \lambda\left(Y_1^a(\bar{e}) - W_0^A - \bar{c}\right)$. In this case, as well, the agent will choose to exert effort at the first opportunity to obtain a precursor. Thus, in both scenarios, the agent will exert effort $\bar{e}$ in the first opportunity.

Conversely, assume that the agent exerts effort at the level of $\bar{e}$ in the first opportunity to try and obtain a precursor. In this case, similarly, consider the scenarios where $W_1^A - \bar{c} \geq W_0^A$ and $W_1^A - \bar{c} < W_0^A$. First, in the case of $W_1^A - \bar{c} \geq W_0^A$, according to Lemma 5.5, $Y_1^a(\bar{e}) - \bar{c} \geq W_1^A - \bar{c} \geq W_0^A$. On the contrary, in the case $W_1^A - \bar{c} < W_0^A$, Lemma 5.6 implies that if the first effort opportunity is avoided, the second will certainly be avoided. Therefore, when choosing to exert effort at the first opportunity, it must be that $Y_1^a(\bar{e}) - W_0^A \geq \bar{c}$. Hence, the necessary and sufficient condition for exerting effort to obtain a precursor in the first trial is $Y_1^a(\bar{e}) - \bar{c} \geq W_0^A$.

When $D^A \in (D_+^M, D_1^A)$,

$$
\begin{aligned}
&\frac{d}{dD^A}\left(Y_1^a(\bar{e}) - W_0^A\right)\\
&= \frac{d}{dD^A}E_{\boldsymbol{x}_1}\left(\max\left(W_1^a(x_1), W_1^a(x_1) + \lambda\left(W_2^a(x_1) - W_1^a(x_1) - \bar{c}\right)\right)|\boldsymbol{x}_0\right) - \frac{dW_0^A}{dD^A}\\
&= (1-\tau) - \int_0^{\chi_1}\frac{\partial W_1^a(x_1)}{\partial D^A}\log(x_1)dx_1\\
&\quad - \int_{\chi_1}^{x_1^*}\left(\lambda\frac{\partial W_2^a(x_1)}{\partial D^A} + (1-\lambda)\frac{\partial W_1^a(x_1)}{\partial D^A}\right)\log(x_1)dx_1\\
&\quad - \int_{x_1^*}^{x_2^*}\lambda\frac{\partial W_2^a(x_1)}{\partial D^A}\log(x_1)dx_1\\
&= (1-\tau) - \int_0^{x_1^*}\frac{\partial W_1^a(x_1)}{\partial D^A}\log(x_1)dx_1 - \lambda\int_{x_1^*}^{x_2^*}\frac{\partial W_2^a(x_1)}{\partial D^A}\log(x_1)dx_1\\
&\quad -\lambda\int_{\chi_1}^{x_1^*}\left(\frac{\partial W_2^a(x_1)}{\partial D^A} - \frac{\partial W_1^a(x_1)}{\partial D^A}\right)\log(x_1)dx_1\\
&> (1-\tau) + \int_0^{\tau^{D^M/B}}\frac{\log(\tau)}{\log(x_1)}\log(x_1)dx_1 - \lambda\int_{\tau^{D^M/B}}^{\delta\tau}\frac{\delta(1-\tau)}{\log(x_1)}\log(x_1)dx_1\\
&\quad \left(\because\ \frac{\partial W_1^a(x_1)}{\partial D^A} = -\frac{\log(\tau)}{\log(x_1)} < \frac{\delta(1-\tau)}{\log(x_1)} = \frac{\partial W_2^a(x_1)}{\partial D^A} < 0,\quad \text{for } x_1 \in (0, x_1^*)\right)\\
&= (1-\tau) + \log(\tau)\tau^{D^M/B} - \lambda\delta(1-\tau)(\delta\tau - \tau^{D^M/B})\\
&> (1-\tau) + \delta\tau\left(1 - \frac{1}{\tau}\right) - \delta(1-\tau)(\delta\tau - \tau^{D^M/B})\\
&\quad \left(\because\ \log(\tau) > 1 - \frac{1}{\tau},\ \text{for } x_1 \in (0,1),\ \text{and } \delta\tau > \tau^{D^M/B}\right)\\
&= (1-\tau)\left(1 - \delta - \delta^2\tau + \delta\tau^{D^M/B}\right) > 0\\
&\quad \Big(\because\ \text{Let } \omega(D^M) \equiv 1 - \delta - \delta^2\tau + \delta\tau^{D^M/B}\\
&\quad \omega(B) = 0\ (\because \delta = 1,\ \text{for } D^M = B),\\
&\quad \omega'(D^M) = (1 + 2\delta\tau - \tau^{D^M/B})\delta^2(1-\tau)/B + \delta\tau^{D^M/B}\log(\tau)/B\\
&\quad > \{\tau\delta(-\delta(1-\tau) + \log(\tau)) + (1+2\delta\tau)\delta(1-\tau)\}\,\delta/B\\
&\quad\ (\because \tau^{D^M/B} < \delta\tau)\\
&\quad > \{\tau\delta(-\delta(1-\tau) + 1 - 1/\tau) + (1+2\delta\tau)\delta(1-\tau)\}\,\delta/B\\
&\quad = \delta^3\tau(1-\tau)/B > 0\Big)
\end{aligned}
$$

When $D^A \in [D_1^A, D_2^A)$,

$$
\begin{aligned}
&\frac{d}{dD^A}\left(Y_1^a(\bar{e}) - W_0^A\right)\\
&= (1-\tau) - \int_0^{\chi_1} \frac{\partial W_1^a(x_1)}{\partial D^A}\log(x_1)dx_1 - \int_{x_1^*}^{\chi_2} \lambda \frac{\partial W_2^a(x_1)}{\partial D^A}\log(x_1)dx_1\\
&\quad - \int_{\chi_1}^{x_1^*}\left(\lambda \frac{\partial W_2^a(x_1)}{\partial D^A} + (1-\lambda)\frac{\partial W_1^a(x_1)}{\partial D^A}\right)\log(x_1)dx_1\\
&= (1-\tau) - \int_0^{x_1^*} \frac{\partial W_1^a(x_1)}{\partial D^A}\log(x_1)dx_1 - \lambda\int_{x_1^*}^{\chi_2} \frac{\partial W_2^a(x_1)}{\partial D^A}\log(x_1)dx_1\\
&\quad -\lambda \int_{\chi_1}^{x_1^*}\left(\frac{\partial W_2^a(x_1)}{\partial D^A} - \frac{\partial W_1^a(x_1)}{\partial D^A}\right)\log(x_1)dx_1\\
&> (1-\tau) + \int_0^{\tau^{D^M/B}} \frac{\log(\tau)}{\log(x_1)}\log(x_1)dx_1\\
&\quad -\lambda \int_{\tau^{D^M/B}}^{\delta\tau} \frac{\delta(1-\tau)}{\log(x_1)}\log(x_1)dx_1 > 0\\
&\quad (\because\ \chi_2 < x_2^*)
\end{aligned}
$$

When $D^A \in [D_2^A, \infty)$,

$$
\begin{aligned}
&\frac{d}{dD^A}\left(Y_1^a(\bar{e}) - W_0^A\right)\\
&= (1-\tau) - \int_0^{\chi_1} \frac{\partial W_1^a(x_1)}{\partial D^A}\log(x_1)dx_1\\
&\quad - \int_{\chi_1}^{x_1^*}\left(\lambda \frac{\partial W_2^a(x_1)}{\partial D^A} + (1-\lambda)\frac{\partial W_1^a(x_1)}{\partial D^A}\right)\log(x_1)dx_1\\
&= (1-\tau) - \int_0^{x_1^*} \frac{\partial W_1^a(x_1)}{\partial D^A}\log(x_1)dx_1\\
&\quad -\lambda \int_{\chi_1}^{x_1^*}\left(\frac{\partial W_2^a(x_1)}{\partial D^A} - \frac{\partial W_1^a(x_1)}{\partial D^A}\right)\log(x_1)dx_1\\
&> (1-\tau) + \int_0^{\tau^{D^M/B}} \frac{\log(\tau)}{\log(x_1)}\log(x_1)dx_1\\
&= 1-\tau+\tau^{D^M/B}\log(\tau) > 1-\tau+\tau\log(\tau) > 0.
\end{aligned}
$$

□

Proof of Lemma 6.1 The subjective posterior probability of an accident occurring for the agent given a value of precursor x_1 is shared with the DM, as shown in Lemma 3.1. The project value is optimized point-wisely for the agent if the project is executed only when $B + R - D^A p(\theta > \tau | x_1) \geq 0$. □

Proof of Lemma 6.3 For Cases 1 and 4, as $H_{\sigma_0^*}(\boldsymbol{x}_0) = 0$ the project is not executed under the decision σ_0^*, and then $W(\boldsymbol{x}_0; \sigma_0^*, B', D') = 0, \forall B', \forall D'$. Furthermore, as $\rho = \sigma_1^* \cap \sigma_1^A$, $\rho = \sigma_1^*$ for Case 1, and $\rho = \sigma_1^A$ for Case 4. For Cases 2 and 3, as $H_{\sigma_0^*}(\boldsymbol{x}_0) = 1$ the project is executed under the decision σ_0^*, and then, the project values are $W(\boldsymbol{x}_0; \sigma_0^*, B, D^M) = W_0^M$ and $W(\boldsymbol{x}_0; \sigma_0^*, B + R, D^A) = W_0^A$. As $\rho = \sigma_1^* \cup \sigma_1^A$, $\rho = \sigma_1^*$ for Case 2 and $\rho = \sigma_1^A$ for Case 3. □

Proof of Lemma 6.4 For any function $g(x)$ on $x \in \Sigma$,

$$\max_{\sigma} \int_{\sigma \subset \Sigma} g(x)dx = \int_{\{x|g(x) \geq 0, x \in \Sigma\}} g(x)dx.$$

Then,

$$\max_{\sigma_1} \Phi(B + R, D^A, \sigma_1) = \max_{\sigma_1} \int_{\sigma_1} \left\{ (B + R) - D^A \frac{\log(\tau)}{\log(x)} \right\} (-\log(x))dx$$

$$= \int_{\sigma_1^A} \left\{ (B + R) - D_A \frac{\log(\tau)}{\log(x)} \right\} (-\log(x))dx$$

$$= \int_0^{\tau^{\frac{D_A}{B+R}}} \{-(B + R)\log(x) + D_A \log(\tau)\} dx = (B + R)\tau^{\frac{D_A}{B+R}}$$

The same procedure is applied for $\Phi(B, D^M, \sigma_0^*)$. □

Proof of Proposition 6.1 As $D_A > \bar{D}_A^*$, $W_1^A - W_0^A > 0$ by Lemma 4.2 and $e^* \in (0, \bar{e})$ exists by Proposition 4.1. The condition (4.3) for e^* gives

$$W_1^A - W_0^A = \frac{c'(e^*)}{\pi'(e^*)}.$$

Differentiating the right-hand side of the condition gives

$$\frac{d}{de^*} \frac{c'(e^*)}{\pi'(e^*)} = \frac{c''(e^*)\pi'(e^*) - c'(e^*)\pi''(e^*)}{\{\pi'(e^*)\}^2} > 0.$$

For each case i $(i = 1 \cdots 4)$, e_i satisfies the condition

$$\frac{d}{de} V_i^A - c'(e) = 0$$

by Definition 6.6. When $W_0^A > 0$, these conditions yield

$$\frac{c'(e)}{\pi'(e)} = \begin{cases} \Phi(B+R, D^A, \sigma_1^*) = W_1^A > W_1^A - W_0^A & Case\ 1 \\ \Phi(B+R, D^A, \sigma_1^*) - W_0^A = W_1^A - W_0^A & Case\ 2 \\ \Phi(B+R, D^A, \sigma_1^A) - W_0^A \geq W_1^A - W_0^A & Case\ 3 \\ \Phi(B+R, D^A, \sigma_1^A) \geq W_1^A > W_1^A - W_0^A & Case\ 4 \end{cases} \tag{A.4}$$

Then, for each case,

$$\frac{c'(e_i)}{\pi'(e_i)} \geq \frac{c'(e^*)}{\pi'(e^*)}.$$

The negative value of W_0^A affects the size relations (A.4) only for Cases 1 and 4. Especially for Case 1, if $W_0^A < 0$, then $\Phi(B+R, D_A, \sigma_1^*) = W_1^A < W_1^A - W_0^A$. □

Proof of Lemma 6.5 For Cases 1 and 3, the first-order conditions to determine the effort levels are

$$\begin{aligned} &\Phi(B+R, D^A, \sigma_1^*) \\ &= \tau^{\frac{D^M}{B}} \left\{ B+R - \left(\frac{B+R}{B} D^M - D^A \right) \log(\tau) \right\} = \frac{c'(e_1)}{\pi'(e_1)} \end{aligned} \tag{A.5}$$

$$\begin{aligned} &\Phi(B+R, D^A, \sigma_1^A) - W_0^A \\ &= (B+R)\tau^{\frac{D^A}{B+R}} - \left\{ (B+R) - (1-\tau)D^A \right\} = \frac{c'(e_3)}{\pi'(e_3)} \end{aligned} \tag{A.6}$$

respectively. Assume $D^A \in (B+R, (B+R)D^M/B]$. Denote the left-hand side of Eqs. (A.5) and (A.6) as function $\phi_0(D^A)$ and $\phi_1(D^A)$, respectively. For $D^A \leq (B+R)D^M/B$, $\phi_0(D^A) > 0$, and then, $e_1 > 0$. Because

$$\begin{aligned} \phi_1(B+R) &= (B+R)(1-\tau) - (B+R)(1-\tau) = 0, \\ \phi_1'(D^A) &= (1-\tau) + \tau^{\frac{D^A}{B+R}} \log(\tau) > 1 - \tau + \tau \log(\tau) > 0, \end{aligned}$$

$\phi_1(D^A) > 0$, so that $e^3 > 0$. Define $\phi_2(D^A) \equiv \phi_0(D^A) - \phi_1(D^A)$. Then,

$$\begin{aligned} \phi_2\left(\frac{B+R}{B} D^M\right) &= \frac{B+R}{B}(B - (1-\tau)D^M) > 0 \\ \phi_2'(D_A) &= \tau^{\frac{D^M}{B}} \log(\tau) - \left\{ \tau^{\frac{D^A}{B+R}} \log(\tau) + (1-\tau) \right\} \\ &< \tau^{\frac{D^M}{B}} \log(\tau) - \{ \tau \log(\tau) + (1-\tau) \} < 0. \end{aligned}$$

Thus, $\phi_2(D^A) > 0$ for $D^A \leq (B+R)D^M/B$, so that $e_1 > e_3$.

Next, assume $D^A \leq B+R$. As $D^A \leq B+R < (B+R)D^M/B$, $e_1 > 0$. $\phi_1(D^A) \leq 0$ for $D^A \leq B+R$, and then, $e_3 = 0$. □

Proof of Proposition 6.2 First, consider the case $\tau \to (1 - B/D^M)^+$. From Lemma 6.3,

$$\lim_{\tau \to \tau_0^+} V_1^M = \pi(e_1) \cdot B\tau_0^{\frac{D^M}{B}} \tag{A.7}$$

$$\lim_{\tau \to \tau_0^+} V_3^M = \pi(e_3) \cdot \tau_0^{\frac{D^A}{B+R}} \left\{ B - \left(\frac{B}{B+R} D^A - D^M \right) \log(\tau_0) \right\} \tag{A.8}$$

where $\tau_0 \equiv 1 - B/D^M$.

Consider the subcase $D^A \in (B + R, (B + R)D^M/B]$. By Lemma 6.5, $e_1 > 0$ and $e_3 > 0$. Because, Eq. (A.7) has a positive value and as

$$B - \left(\frac{B}{B+R} D^A - D^M \right) \log(\tau_0) > B - (B - D^M) \log(\tau_0)$$
$$= B\left(1 - \left(1 - \frac{D^M}{B}\right) \log\left(1 - \frac{B}{D^M}\right)\right) > 0 \quad \text{for } D^M > B,$$

Equation (A.8) has also a positive value. Note that for the function $\phi_3(x) = 1 - (1 - x)\log(1 - 1/x)$, $\phi_3(x) > 0$ for $x > 1$, because $\phi_3'(x) = \log(1 - 1/x) - (1 - x)/(x^2(1 - 1/x)) < 0$ for $x > 1$ and $\lim_{x\to\infty} \phi_3(x) = 0$. Here, the ratio of values of Eqs. (A.7) and (A.8) is

$$\frac{\pi(e_3)}{\pi(e_1)} \cdot \frac{\tau_0^{\frac{D^A}{B+R}}}{\tau_0^{\frac{D^M}{B}}} \left(1 - \log\left(\frac{\tau_0^{\frac{D^A}{B+R}}}{\tau_0^{\frac{D^M}{B}}} \right) \right).$$

By Lemma 6.5, $e_1 > e_3$, which leads to $1 > \pi(e_3)/\pi(e_1)$. Moreover, when $1 - \log(x) > 0$, $x(1 - \log(x)) \le 1$ for $x \in \mathbb{R}^{++}$. Thus, the ratio takes the value of less than 1. Then, the DM chooses Case 1, $H_{\sigma_0^*}(\boldsymbol{x}_0) = 0$.

Next, consider the subcase $D^A \in (0, B + R]$. By Lemma 6.5, $e_1 > e_3 = 0$. Then, Eq. (A.7) takes a positive value, and the value of Eq. (A.8) is 0. Then, the DM chooses Case 1 again.

In the same manner, when $\tau \to 1^-$,

$$\lim_{\tau \to 1^-} V_1^M = \pi(e_1) \cdot B < B = \pi(e_3) \cdot B + (1 - \pi(e_3)) \cdot B = \lim_{\tau \to 1^-} V_3^M.$$

Then, the DM chooses Case 3, $H_{\sigma_0^*}(\boldsymbol{x}_0) = 1$. □

Proof of Proposition 6.3 Here, Eqs. (A.5), (A.6), and the functions $\phi_0(\cdot)$, $\phi_1(\cdot)$ are those defined in Proof of Lemma 6.5. e_1 and e_3 satisfy conditions (A.5) and (A.6), respectively. Then, by the implicit function theorem,

$$
\begin{aligned}
\mathrm{sign}\left(\frac{de_1}{dD^A}\right) &= \mathrm{sign}\left(\frac{1}{\frac{\partial \phi_0}{\partial D^A}}\left(\frac{c''\pi' - c'\pi''}{(\pi')^2}\right)\right) \\
&= \mathrm{sign}\left(\frac{\partial \phi_0}{\partial D^A}\right) = \mathrm{sign}\left(\tau^{\frac{D^M}{B}}\log(\tau)\right) = -1 \\
\mathrm{sign}\left(\frac{de_1}{dR}\right) &= \mathrm{sign}\left(\frac{\partial \phi_0}{\partial R}\right) = \mathrm{sign}\left(\tau^{\frac{D^M}{B}}\left(1 - \frac{D^M}{B}\log(\tau)\right)\right) = +1 \\
\mathrm{sign}\left(\frac{de_3}{dD^A}\right) &= \mathrm{sign}\left(\frac{\partial \phi_1}{\partial D^A}\right) = \mathrm{sign}\left(\tau^{\frac{D^A}{B+R}}\log(\tau) + 1 - \tau\right) \\
&\geq \mathrm{sign}\left(\tau\log(\tau) + 1 - \tau\right) = +1 \text{ for } D^A > B + R, \\
\mathrm{sign}\left(\frac{de_3}{dR}\right) &= \mathrm{sign}\left(\frac{\partial \phi_1}{\partial R}\right) = \mathrm{sign}\left(\tau^{\frac{D^A}{B+R}} - 1 - \tau^{\frac{D^A}{B+R}}\log\left(\tau^{\frac{D^A}{B+R}}\right)\right) = -1
\end{aligned}
$$

Note that $\forall x \in \mathbb{R}^{++},\ x \neq 1,\ x - 1 - x\log(x) < 0$, and $B + R < D^A$ when $e_3 > 0$ by Lemma 6.5. □

Proof of Lemma 6.6 For Cases 2 and 4, as shown in Proof of Proposition 6.1 the conditions that determine the effort of the agent are

$$
\begin{aligned}
&\Phi(B+R, D^A, \sigma_1^*) - W_0^A \\
&= \tau^{\frac{D^M}{B}}\left(B + R - \left(\frac{B+R}{B}D^M - D^A\right)\log(\tau)\right) - \left(B + R - (1-\tau)D^A\right) \\
&= (B+R)\left(\tau^{\frac{D^M}{B}} - 1 - \tau^{\frac{D^M}{B}}\log(\tau^{\frac{D^M}{B}})\right) + D^A(1 - \tau + \tau^{\frac{D^M}{B}}\log(\tau)) \\
&= \frac{c'(e_2)}{\pi'(e_2)}
\end{aligned}
\tag{A.9}
$$

$$
\Phi(B+R, D^A, \sigma_1^A) = (B+R)\tau^{\frac{D^A}{B+R}} = \frac{c'(e_4)}{\pi'(e_4)} \tag{A.10}
$$

respectively. Here, when $D^A > (B+R)D^M/B$, the left-hand side of Eq. (A.9) takes a positive value because

$$
\begin{aligned}
&(B+R)\left(\tau^{\frac{D^M}{B}} - 1 - \tau^{\frac{D^M}{B}}\log(\tau^{\frac{D^M}{B}})\right) + D^A(1 - \tau + \tau^{\frac{D^M}{B}}\log(\tau)) \\
&> \frac{BD^A}{D^M}\left(\tau^{\frac{D^M}{B}} - 1 - \tau^{\frac{D^M}{B}}\log(\tau^{\frac{D^M}{B}})\right) + D^A(1 - \tau + \tau^{\frac{D^M}{B}}\log(\tau)) \\
&(\because x - 1 - x\cdot\log(x) < 0,\ \text{for } x \in (0,1)) \\
&= D^A\left(1 - \tau + \frac{B}{D^M}\left(\tau^{\frac{D^M}{B}} - 1\right)\right) > 0 \\
&(\because 1 - x + (x^y - 1)/y > 0,\ \text{for } y > 1 \text{ and } x \in (0,1)).
\end{aligned}
$$

Then, $e_2 > 0$. The left-hand side of Eq. (A.10) also, evidently, takes a positive value, then $e_4 > 0$. Denote the difference between both equations, $(\Phi(B+R, D^A, \sigma_1^*) - W_0^A) - \Phi(B+R, D^A, \sigma_1^A)$ as a function $\phi_4(D^A)$. Then,

$$\begin{aligned}
&\phi_4\left(\frac{B+R}{B}D^M\right) = -\frac{B+R}{B}(B-(1-\tau)D^M) < 0\\
&\phi_4\left(\frac{B+R}{1-\tau}\right) =\\
&(B+R)\left(\tau^{\frac{D^M}{B}} - \tau^{\frac{1}{1-\tau}} - \tau^{\frac{D^M}{B}}\left(\log\left(\tau^{\frac{D^M}{B}}\right) - \log\left(\tau^{\frac{1}{1-\tau}}\right)\right)\right) < 0\\
&(\because x - a - x(\log(x) - \log(a)) < 0, \ \text{for } 0 < a < x < 1)\\
&\phi_4'(D^A) = \left(\tau^{\frac{D^M}{B}} - \tau^{\frac{D^A}{B+R}}\right)\log(\tau) + (1-\tau)\\
&> \tau^{\frac{D^M}{B}}\log(\tau) + (1-\tau) > \tau\log(\tau) + 1 - \tau > 0.
\end{aligned}$$

Thus, $\phi_4(D^A) < 0, \forall D^A \in (D^M(B+R)/B, (B+R)/(1-\tau))$. This leads $e_2 < e_4$. Furthermore,

$$W_0^A = B + R - (1-\tau)D^A > 0 \Leftrightarrow D^A < \frac{B+R}{1-\tau}.$$

As

$$\frac{B+R}{1-\tau} - \frac{B+R}{B}D^M = \frac{B+R}{B(1-\tau)}(B-(1-\tau)D^M) > 0$$

the region $D^A \in (D^M(B+R)/B, (B+R)/(1-\tau))$ is not empty. Furthermore,

$$\lim_{D^A\to\infty}\frac{\phi_4(D^A)}{D^A} = \tau^{\frac{D^M}{B}}\log(\tau) + 1 - \tau > 0,$$

leads $\exists\tilde{D}^A, \ \phi_4(D^A) \geq 0, \ \ \forall D^A \geq \tilde{D}^A$, which implies $e^4 < e^2$. □

Poof of Proposition 6.4

$$\begin{aligned}
V_2^M &= (1-\pi(e))\,W_0^M + \pi(e)\Phi(B, D^M, \sigma_1^*)\\
&= (1-\pi(e))\,W_0^M + \int_0^{\tau^{\frac{D^M}{B}}}\left(-B\log(x_1)dx_1 + D^M\log(\tau)\right)dx_1\\
&= (1-\pi(e))\left(B-(1-\tau)D^M\right) + B\tau^{\frac{D^M}{B}},
\end{aligned}$$

$$
\begin{aligned}
V_4^M &= \pi(e)\Phi(B, D^M, \sigma_1^A) \\
&= \pi(e)\int_0^{\tau^{\frac{D^A}{B+R}}} \left(-B\log(x_1)dx_1 + D^M\log(\tau)\right)dx_1 \\
&= \tau^{\frac{D^A}{B+R}}\left(B + \left(D^M - \frac{B}{B+R}D^A\right)\log(\tau)\right).
\end{aligned}
$$

As $\pi(e_4) < 1$,

$$
\lim_{\tau\to 1^-} V_4^M = \pi(e_4)\cdot B < B = \pi(e_2)\cdot B + (1-\pi(e_2))\cdot B = \lim_{\tau\to 1^-} V_2^M,
$$

when $\tau \to 1^-$, the DM chooses Case 2, where $H_{\sigma_0^*}(\boldsymbol{x}_0) = 1$.

Denote τ_0 as $\tau_0 \equiv 1 - B/D^M$. As

$$
\lim_{\substack{\tau\to\tau_0^+ \\ D^A\to\frac{B+R}{B}D^M}} V_2^M = \pi(e_2)\cdot B\tau^{\frac{D^M}{B}}, \qquad \lim_{\substack{\tau\to\tau_0^+ \\ D^A\to\frac{B+R}{B}D^M}} V_4^M = \pi(e_4)\cdot B\tau^{\frac{D^M}{B}}
$$

and $\pi(e_2) < \pi(e_4)$ when $D^A \to D^M(B+R)/B^+$ from Lemma 6.6, $\pi(e_2) < \pi(e_4)$. Therefore, when $\tau \to \tau_0^+$ and $D^A \to D^M(B+R)/B$, DM chooses Case 4, where $H_{\sigma_0^*}(\boldsymbol{x}_0) = 0$. □

Proof of Proposition 6.5 Denote the left-hand sides of Eqs. (A.9) and (A.10) as $\phi_5(D^A, R)$ and $\phi_6(D^A, R)$, respectively. In the same manner as the proof of Proposition 6.3,

$$
\begin{aligned}
\text{sign}\left(\frac{de_2}{dD^A}\right) &= \text{sign}\left(\frac{\partial\phi_5}{\partial D^A}\right) = \text{sign}\left(\tau^{\frac{D^M}{B}}\log(\tau) + 1 - \tau\right) \\
&\geq \text{sign}\left(\tau\log(\tau) + 1 - \tau\right) = +1, \\
\text{sign}\left(\frac{de_2}{dR}\right) &= \text{sign}\left(\frac{\partial\phi_5}{\partial R}\right) = \text{sign}\left(\tau^{\frac{D^M}{B}}\left(1 - \frac{D^M}{B}\log(\tau)\right) - 1\right) \\
&= \text{sign}\left(\tau^{\frac{D^M}{B}}\left(1 - \log(\tau^{\frac{D^M}{B}})\right) - 1\right) = -1, \\
&(\because x(1-\log(x)) - 1 < 0, \text{ for } x \in (0,1)) \\
\text{sign}\left(\frac{de_4}{dD^A}\right) &= \text{sign}\left(\frac{\partial\phi_6}{\partial D^A}\right) = \text{sign}\left(\tau^{\frac{D^A}{B+R}}\log(\tau)\right) = -1, \\
\text{sign}\left(\frac{de_4}{dR}\right) &= \text{sign}\left(\frac{\partial\phi_6}{\partial R}\right) = \text{sign}\left(\tau^{\frac{D^A}{B+R}}\left(1 - \log\left(\tau^{\frac{D^A}{B+R}}\right)\right)\right) = +1.
\end{aligned}
$$

□

Part II
Trial-Error Decision-Making and Severe Accident

Chapter 8
Introduction: Purpose of Part 2 and Background of the Issue

8.1 Normal Accident and High Reliable Organization

Charles Perrow's research on "normal accidents" is arguably the most influential analysis concerning severe accidents and their preventability. Perrow conducted an in-depth analysis of the 1979 accident at the Three Mile Island nuclear power plant in the United States, characterizing it as a "normal accident"—an inevitable event. He argued that "normal" or system accidents are inherent to systems characterized by complexity and tight coupling (Perrow 1984, 1999).

Complexity, one of the system characteristics identified by Perrow, refers to a system composed of numerous elements that interact with each other. This contrasts with linear systems, where each element is sequentially and unidirectionally related, akin to items on a conveyor belt. In a complex system, elements can be adjacent or become sources of unexpected effects, leading to interference between elements and feedback loops. Understanding the entire system is necessary to grasp all these aspects. However, due to its complexity, full comprehension is challenging, making it difficult to predict the system's overall behavior.

Tight coupling, the other characteristic, refers to the extent to which each system element defines its functions and coordination methods with others. Perrow noted that when the coupling is tight, the coordination between sequentially occurring elements must be precise, resulting in high time dependence on the system and the timing of each element's operation becoming critical to the system's function. Furthermore, there is little flexibility in the sequence of functions, a strict order is required, and there is no interchangeability. When values need to be transmitted within the system, they must be maintained accurately without any margin for error. If a system component malfunctions, tight coupling leaves no room to correct deviations from planned operations, leading to a cumulative expansion of the impact. Conversely, loose coupling provides leeway in coordination and sequence, allowing for flexibility. Therefore, in the event of an external shock, the impact can be limited, and the system is unlikely to become unstable. Perrow posited that the two characteristics

A. Torii, *Decision-making in High-Risk Projects*,
https://doi.org/10.1007/978-981-97-3056-8_8

of complexity and tight coupling are independent of each other, and there can be systems that are linear yet tightly coupled, or complex yet loosely coupled (Perrow 1984, p. 89).

Perrow's detailed analysis of the Three Mile Island nuclear power plant accident demonstrated that the power station, as a system, had high complexity and tight coupling. He explained in detail, with reference to the progression of the accident, that these factors not only caused the accident but also led to its severity. In this context, the elements constituting the system are not just the nuclear reactors and associated equipment and structures. Personnel are also considered as components of the system. Therefore, the organization of the power plant is a major factor in determining the degree of complexity and tight coupling. In Perrow's 1984 work, he analyzed not only the Three Mile Island nuclear power plant accident but also various accidents including petrochemical plant accidents, aviation and maritime disasters. Particularly in the analysis of accidents in space exploration, crew members (personnel) are considered as one of the components necessary for the system's operation and are analyzed as factors leading to the occurrence of accidents.

Following the publication of Perrow's work in 1984, a series of significant disasters, termed severe accidents, occurred, almost as if predicted by his theories. These include the explosion of the U.S. Space Shuttle Challenger and the Chernobyl nuclear power plant accident in the former Soviet Union. These accidents predominantly occurred in the latter half of the 20th century.[1] In the context of these severe accidents, the notion of understanding these incidents as normal accidents gained prominence.[2] It is crucial to understand that the complexity and tight coupling discussed by Perrow refer to the characteristics of the entire nuclear power plant system, including personnel. The analysis targets the technical and engineering aspects of the system. For instance, nuclear power generation involves the transformation of materials, inherently making the sequence and timing of processes important, leading to tight coupling.

Subsequent to Perrow's work, a significant discussion emerged about High-Reliability Organization (HRO). Roberts (1990a; 1990b) used the example of the landing and takeoff systems on U.S. Navy nuclear-powered aircraft carriers to demonstrate that organizations can maintain high reliability over extended periods without accidents, even in very complex and high-risk systems. Other organizations considered as HRO include power companies' electricity transmission networks, air traffic control systems, and nuclear power plants. These organizations have been analyzed to determine the necessary conditions for constructing an organization that suppresses the occurrence of accidents. For instance, the clarification of safety objectives and the establishment of a learning system from mistakes are deemed necessary.

In this context, HRO are positioned as a counterargument to the concept of normal accidents (Roberts 1990a). The argument posits that systems, even if they are complex and tightly coupled, are not necessarily vulnerable to accidents if there is an appropriate organizational response. While Perrow's discussion of complexity and

[1] Roberts (1990b).

[2] For instance, see Pidgeon (2011).

tight coupling tends to focus more on the technical nature of the system's equipment and structures, research on HRO seeks to clarify the nature of human organizations that prevent accidents.

In his second edition, Perrow (1999), commenting on the discussion of HRO and referencing Sagan (1993), explains that the characteristics attributed to HRO are more effective in reducing risks faced by such organizations when the coupling is loose. He argues that even the aspect of learning, which researchers of HRO deem most crucial, becomes extremely difficult in situations of tight coupling. This difficulty arises because dysfunctions interfere and intertwine in unexpected and incomprehensible ways, making it hard to accumulate experience and learn from it.

In the discussion of HRO, redundancy is considered a crucial response to the two system factors of complexity and tight coupling. Redundancy refers to multiple layers of monitoring and safety measures. In Roberts (1990a), it is explained how in electricity transmission system management and air traffic control systems, multiple power dispatch operators and controllers back each other up and share operational duties to handle complex and continuous events. In nuclear power generation as well, safety measures against accidents are set up in multiple layers. If these function as backups, in theory, the probability of a major accident should be kept very low.

Regarding redundancy's effectiveness, Perrow (1984) uses the U.S. space program as an example to argue that increased redundancy, demanded by a system requiring more flexibility, actually led to greater complexity and unexpected behaviors. Additionally, Sagan (2004) provides three reasons why redundancy may not function in normal accidents. First, he discusses the possibility of common-mode errors, where multiple safety measures fail simultaneously during an accident, rendering backup measures ineffective. He notes that in nuclear power plants, safety devices often cause the most accidents ("safety device accidents"). Second, individuals tend to neglect their responsibilities when they know others can fulfill those duties ("social shirking"). Lastly, efforts to maintain safety tend to decrease when safety-enhancing measures are implemented ("overcompensation"), a moral hazard problem. The latter two reasons pertain to human organization issues.

Researchers studying HRO highlight that these organizations possess various risk-reducing characteristics in addition to redundancy. They propose numerous operational requirements, such as "learning from failures," to mitigate accidents.[3] In nuclear power generation, a comprehensive examination and evaluation of these organizational operations' possibilities would be beneficial. This task necessitates a highly technical systems analysis. To ascertain whether the human systems organized for operation effectively prevent severe accidents and identify improvement areas, numerous case studies are required.

The purpose of this part of the book is not to investigate the feasibility of constructing a system that avoids accidents using a specific project as an example. Instead, the focus is on whether the design of decision-making systems in projects with severe accident potential can prevent such accidents. The question is what challenges exist

[3] For an overview, refer to works such as Weick and Sutcliffe (2001), Nishimoto (2006), and Takagi (2006).

in designing decision-making systems for such projects. More specifically, during the research and development process, which explores the level and nature of risks while trialing the project, the issue is identifying the conditions and challenging factors of a decision-making system that can maximally prevent severe accidents when signs indicating a severe accident's risk nature are obtained and used for informed decision-making. In the remaining part of this chapter, using nuclear power generation as an archetype of a project with complexity and tight coupling characteristics, the significance and background of specifying and analyzing the problem as described above will be explained.

If nuclear power generation technology possesses the two characteristics of complexity and tight coupling, does this mean it is inevitably vulnerable to accidents even during the development process deemed necessary for the future? If there is a possibility to avoid this vulnerability, how should the system be designed?

8.2 Research and Development in the Nuclear Power Business

Perrow explains that the complexity and tight coupling of the nuclear power generation system stem from its development history. After World War II, amid the nuclear arms race, the U.S. government urgently sought to demonstrate the potential for peaceful uses of nuclear power. This urgency drove power companies to develop nuclear power generation. The prototype for commercial reactors at that time was the nuclear reactor developed for military use in nuclear submarines. These submarines required small, highly operable reactors with easy refueling. However, power companies had different specifications. They did not need small reactors, and as a base-load power source, they did not require high operability or easy refueling. Despite these differences, submarine reactors were quickly modified and proportionally enlarged to develop commercial reactors. Perrow surmises that this process led to the design becoming complex and tightly coupled. Martin (2012) also describes this development process, stating that light-water reactors, which use high-pressure water for underwater cooling convenience, were chosen, necessitating a pressure vessel and complex piping.

Perrow does not believe that it is impossible to design safer reactors from the outset. However, he argues that given the number of reactors developed in this way currently in operation, it will be challenging to change the design trend in the coming decades. This view was expressed at the time of the publication of Perrow (1984). Since then, there have been suggestions that the development of nuclear power has essentially stalled since the Three Mile Island nuclear power plant accident, and it is believed that the situation has not changed much even now. Martin (2012) points out that innovation in nuclear technology effectively halted after the 1979 Three Mile Island nuclear plant accident; much of the electricity currently generated by nuclear power comes from older-generation reactors. To gain approval and build new reactors, it is necessary to choose reactor designs that are safe and not excessively innovative, and substantial funding is required. The development delays are indeed a consequence of the Three Mile Island and Chernobyl nuclear plant accidents. Yamaji (2009) notes that the stagnation in nuclear development began even before the Three Mile Island accident, and both incidents exacerbated and prolonged the stagnation.

If the current reactor types were selected for technological necessity or through an evolutionary selection process from a variety of options, the present reactors, especially those of the second generation, which are complex and tightly coupled systems, might be considered technically standardized and matured, leaving little room for further research and development. However, if the current reactor types were chosen for reasons other than technological ones, as Perrow and Martin suggest, they might not be matured but rather subjects that still hold possibilities for modification and research. Perrow notes that despite over a decade of experience in constructing large light-water reactors, significant specifications and design changes are still being made, which is unusual compared to other complex industrial products.

Even so, currently, improved light-water reactors, considered third-generation nuclear reactors, have been developed, with some already under construction and operation. Conversely, fourth-generation nuclear reactors, which fundamentally differ in design from the second-generation reactors built up to the 1990s and are developed from the outset with safety assurance as a design principle for power generation reactors, are just beginning their development. Thus, the reality is that continuous research and development are necessary, and the technology is not yet at a stage where it can be considered mature.[4]

In that case, can we consider that the technology for the construction and operation of existing reactor types, rather than design technology, has matured? The Forum (2005) analyzed the structure of the nuclear technology utilization industry and the nuclear supply industry, which provides equipment and services, through surveys and hearings. Within this analysis, as quoted below, it is noted that tacit knowledge, centered on experience and performance, plays a significant role.

> Moreover, there were many responses related to "experience power." Some companies indicated that "knowledge of unique technologies is core," but generally speaking, in manufacturing, experience in all processes from planning, design, procurement, to assembly is crucial. The responses emphasized the importance of developing and ensuring personnel who have experienced all these stages, who understand nuclear power well, who comprehend the technical background requiring such high quality and management levels in handling nuclear energy, who are organizationally adept in deploying personnel experienced in nuclear operations, who possess "skill quality," and who have "accumulated all experiences." Many companies consider "experience and track record" as the cornerstone of their technological expertise (p. 13).

---

[4] In Japan, the introduction of various types of nuclear reactors as imported technology initially led to low operational rates due to the immaturity of operating techniques, and efficient operation became a challenge. Since the late 1970s, under the concept of "improvement standardization," there has been a focus on improving through the "absorption and assimilation" of technology and on the domestic production of equipment. However, although improvements in equipment utilization rates were observed during this process, the goal of standardization gradually ceased to be pursued. Refer to Ishii (2014) for further details.

> However, there are limits to formalizing tacit knowledge. Tacit knowledge refers to the know-how, instincts, and experiences that become "integral abilities within individuals" through practice and absorption. For instance, skills like "reversing and assembling large containers weighing tens of tons in a factory and then welding them," or "the technique of stellite buildup using electric plasma for metal joining." These skills seem to be efficiently passed on through on-the-job training during the actual process of equipment fabrication and construction (Original note: One manufacturer pointed out that unlike thermal power plants, once assembled, nuclear plants cannot be easily disassembled, and it is hard to see inside. Therefore, accumulating actual experience in design, manufacturing, and installation is crucial to respond to various events) (p. 19).

The importance of tacit knowledge at this stage suggests that there is considerable potential for innovation through the formalization of knowledge in this field. Tacit knowledge is subjective and difficult to express in language, referring to skills embodied through context-dependent experience. Explicit knowledge is objective and can be expressed in language, referring to theoretically organized concepts independent of context. Nonaka and Takeuchi (1996) and Nonaka et al. (1996) describe innovation as a process of knowledge creation, and knowledge creation as a dynamic interaction of conversion between tacit and explicit knowledge.

Furthermore, current nuclear reactor technology and its associated systems are not closed systems. The light-water reactor system is part of the nuclear fuel cycle, which includes the backend processes of reprocessing spent fuel and managing high-level radioactive waste. The maturity of technology related to light-water reactors does not necessarily imply the maturity of the entire nuclear fuel cycle technology. Particularly for the backend, it is assumed that geological disposal will continue for an order of magnitude of years, as indicated by half-lives of tens of thousands of years. If such an extremely long-term project is considered impractical, it becomes necessary at some point to develop group separation and transmutation technologies.[5]

Thus, continuous research and development are still necessary to sustain the nuclear power generation system, and there is significant potential for innovation. With this understanding as a premise, the relationship between normal accidents and HRO takes on a different aspect. HRO are organizations that can sufficiently reduce accident rates through organizational measures such as efficient learning. Therefore, organizations analyzed as HRO may not necessarily achieve sufficiently low accident rates in the initial stages of operation.

In Roberts (1990a), the accident rate of the U.S. Navy's nuclear-powered aircraft carrier takeoff/landing system, often cited as a representative example of HRO, shows a rapid decline as experience accumulates. However, in the early stages where learning was not sufficiently advanced, the accident rate was high.[6] Even if low accident rates can be achieved through various organizational measures, this does not imply that low rates are realized from the beginning of operations. The discussion on HRO does not contradict the concept of normal accidents in this respect. Given the high

---

[5] Yamaji (2009) according to Yamaji, group separation and transmutation refer to the process of separating and recovering long-lived radioactive isotopes from waste and converting them into short half-life isotopes using fast reactors.

[6] Roberts (1990a) p. 103, Fig. 1.

risk in the development stages of nuclear power technology, which is complex and tightly coupled, there is a high possibility of serious accidents occurring before the organization has had the chance to learn, even if the goal is to build an HRO.

Research and development in high-risk systems fundamentally differ from the issues faced in the operation of such systems. In the operation of high-risk systems, as discussed in the context of HRO, the challenge lies in how to learn efficiently and reduce risks during operation. Conversely, in research and development in high-risk systems, the crucial aspect is how to avoid severe accidents during the learning process associated with the system's operation.

Martin (2012) notes that in the development of nuclear power technology, most nuclear engineers are caught in incrementalism, unable to make significant moves. Only procedures with known outcomes are allowed, and the only acceptable risks are those that involve minor changes to existing technologies. This implies that research and development activities in projects considered high risk must choose directions that minimize risk to avoid severe accidents. It is impossible to understand the entire research and development path and choose the one that minimizes risk among all paths, nor is it clear how much risk is associated with each individual path. In such cases, as Martin points out, one is forced to proceed by making choices within a range where risks are somewhat predictable and close at hand. In essence, researchers are forced to conduct research and development by feeling-their-way-forward.

There are several issues with this feeling-their-way-forward or groping-in-the-dark approach to research and development. Groping in the dark is necessary, needless to say, to avoid risks as much as possible. Therefore, the magnitude of the risks must be constantly predicted and assessed, and the progress of the project must be controlled so that it can be stopped if necessary. Analyzing the difficulties in executing this task is the purpose of this thesis.

Remaining chapters in Part 2 explore what HRO means in the context of hazardous research and development activities by modifying and transforming the risk model used in Part 1. Chapter 9 will explain the framework of the model and conduct an analysis using the basic form where learning is not possible. Chapter 10 will analyze the case where learning is possible. Chapter 11, the conclusion, explains that the models presented in the two chapters correspond to tight coupling and loose coupling, respectively.

References

Forum, Japan Atomic Industrial 2005."Current Status and Challenges of Nuclear Supply Industry in Japan." *Nuclear Energy Research Bulletin*. (70): , in Japanese.

Ishii, Susumu 2014."Efficient Nuclear Power Generation and Industrial Policy - Domestic Production and Improvement Standardization." *RIETI Discussion Paper Series*. (14-J-026): , in Japanese.

Martin, Richard 2012. *Superfuel: thorium, the green energy source for the future*. : St. Martin's Press.

Nishimoto, Naoto 2006."Current Status and Issues in HRO Research: Transition of Research Objectives and HRO in Accident Analysis." *Current Status and Challenges of HROs in the ICT Industry, JPCERT/CC*. 1–25, in Japanese.
Nonaka, Ikujiro, and Hirotaka Takeuchi 1996. *Organizational knowledge creation.* : Toyo Keizai Shimp Sha, in Japanese.
Nonaka, Ikujiro, Hirotaka Takeuchi, and Katsuhiro Umemoto 1996."A theory of organizational knowledge creation." *International Journal of Technology Management*. 11 (7-8): 833–845.
Perrow, Charles 1984."Normal accidents: Living with high-risk technologies."
Perrow, Charles 1999. *Normal accidents: Living with high risk technologies.* : Princeton university press.
Pidgeon, Nick 2011."In retrospect: Normal Accidents." *Nature*. 477 (7365): 404–405.
Roberts, Karlene H 1990a."Managing high reliability organizations." *California management review*. 32 (4): 101–113.
Roberts, Karlene H 1990b."Some characteristics of one type of high reliability organization." *Organization science*. 1 (2): 160–176.
Sagan, Scott D 2004."Learning from Normal Accidents." *Organization & Environment*. 17 (1): 15–19.
Sagan, Scott Douglas 1993. *The limits of safety: Organizations, accidents, and nuclear weapons.* : Princeton University Press.
Takagi, Toshio 2006."Possibility of High Reliability Organization Concept and its Empirical Study." *Current Status and Challenges of HROs in the ICT Industry, JPCERT/CC*. 1–9, in Japanese.
Weick, Karl E, and Kathleen M Sutcliffe 2001. *Managing the unexpected.* 9: San Francisco: Jossey-Bass.
Yamaji, Kenji 2009. *The Past, Present, and Future of Nuclear Energy: Will Nuclear Power Be Restored?.* : Corona Pubulishing, in Japanese.

Chapter 9
Model

9.1 Model Framework

As explained in Chap. 8, this chapter analyzes the decision-making process of whether to stop or proceed with a project that involves significant risks. Consider a society comprising a sufficiently large number of different individuals. One of these individuals, referred to as the decision maker (DM hereafter), makes the decision to either halt or continue the project. The execution of the project is entrusted to a single agent, who is a part of the entities constituting the society.

This project is implemented over two periods. Operating this project carries the possibility of an accident. All members of society are aware of the potential for accidents. Unlike the model in Part 1, accidents can occur in the first period of the project, just as in the second period. The project is considered complete and brings a benefit B (where $B > 0$) to society only if it operates in the first and second periods. The DM decides at the beginning of the first and second periods whether to proceed with the project for that period. If the decision to stop the project is made at the beginning of the first period, the project cannot be executed in the second period. That is, the second period of the project can only operate if the first phase is completed. Furthermore, if an accident occurs in the first period, the project cannot be continued into the second period. Additionally, operating only the first period of the project does not generate any benefit. The benefit, resulting from the research and development, is equally distributed among all members of society. There is a social consensus on the value of the benefit B, and all members of society correctly anticipate this value. Unlike the analysis in Part 1, this chapter focuses on the decision to stop the project rather than analyze the incentives for research and development. Therefore, it does not assume any private benefit R arising for the entity implementing the project.

Damage caused by accidents affects all members of society equally as a public good. The damage to society from an accident is represented by a real number D (where $D \geq 0$). No member of society can accurately know the value of D in the first period. There is no information about the probability distribution of D in the

A. Torii, *Decision-making in High-Risk Projects*,
https://doi.org/10.1007/978-981-97-3056-8_9

event of an accident. Each member of society has their own estimate of D and is confident in this value. In particular, let D^M be the DM's estimate of accident damage. Furthermore, in the event of an accident during the execution of a project in the second period, since the project is being carried out throughout the first and second periods, the benefits of the project will occur.

We assume the following is known about the mechanism of accident occurrence:

Assumption 9.8

1. Accidents involve a random variable $X : \Omega \to [0, 1]$ defined in the probability space $(\Omega, \mathcal{F}, P)$.
2. The random variable X follows a continuous uniform distribution $U(0, \theta)$. The value of the distribution parameter θ is unique to the project and satisfies $\theta \in (0, 1]$.
3. In the execution of the project in each period, a value of the random variable X corresponding to an element ω in the set Ω is given. An accident occurs in that period if the value of this random variable X satisfies $X > \tau$. Here, τ is such that $\tau \in (0, 1)$.

When the value of the parameter θ of the random variable X satisfies $\theta > \tau$, an accident occurs in the first and second periods with a constant probability of $(\theta - \tau)/\theta$. However, if $\theta \leq \tau$, as the event where $X > \tau$ does not occur, no accident happens.

The information structure regarding the random variable follows is as follows: Members of society do not know the value of the parameter θ of the random variable X specific to the project. However, the value of the real constant τ is accurately known to all members of society. Assumedly, the DM's estimate of accident damage D^M is made public, and known to all members of society. Furthermore, in this chapter, we analyze the scenario where the realized values of the random variable X in trials are not observable by any member, while in Chap. 10, we examine the case where all members can observe them. The assumption of unobservable values in this chapter is made to consider a state of complete uncertainty and exploration.

Here, we call the prior distribution $F(\theta)$ of the parameter θ, in the absence of any information, a no-information prior distribution. We assume a continuous uniform distribution as the no-information prior distribution.

Assumption 9.9

$$F(\theta) = \begin{cases} 0 \text{ for } \theta \leq 0, \\ \theta \text{ for } 0 < \theta \leq 1, \\ 1 \text{ for } 1 < \theta. \end{cases}$$

Denote the probability density function of this distribution as $f(\theta)$.

9.2 When Learning Is Not Possible

In such projects that proceed through trial and error, knowledge accumulation varies. Some instances allow for knowledge accumulation during the trial-and-error process, while others do not. In the latter case, the trial and error process merely determines whether an accident occurred without distinguishing whether it could have been avoided or not, and to what extent. Projects conducted entirely manually often face decisions in such situations. This chapter first addresses the decision-making challenges in such cases. When limited information is available, a common strategy involves initiating the project temporarily, monitoring its progress, and subsequently assessing the level of risk. The pitfalls associated with this strategy are explored.

Consider the case where none of the members of society can observe the value of the random variable X. That is,

Assumption 9.10 Members of society cannot observe the value of X.

In this case, it is only known that an accident occurs in each period with a constant probability of $\max\left((\theta - \tau)/\theta, 0\right)$.

Under Assumption 9.10, consider the DM's decision-making. First, consider the beginning of the second period. If the project was not carried out in the first period, there is no option to carry out the project in the second period. Moreover, if the project was carried out in the first period and an accident occurred, the project cannot be carried out in the second period. Therefore, consider the case where the project was carried out in the first period and no accident occurred. The only information obtained is the fact that "the project was carried out in the first period and no accident occurred." That is, the only information that can be known is $x_1 \leq \tau$. Denote the probability density function of the posterior distribution of the parameter θ under the condition $x_1 \leq \tau$ as $f(\theta|x_1 \leq \tau)$. Under Assumption 9.9, per Bayes' theorem, we obtain

$$
\begin{aligned}
f(\theta|x_1 \leq \tau) &= \frac{\min\left(\dfrac{\tau}{\theta}, 1\right) f(\theta)}{\displaystyle\int_0^1 \min\left(\dfrac{\tau}{\theta'}, 1\right) f(\theta')d\theta'} \\
&= \begin{cases} \dfrac{1}{\tau(1-\log(\tau))} & \text{when } 0 \leq \theta \leq \tau \\[2ex] \dfrac{1}{\theta(1-\log(\tau))} & \text{when } \tau < \theta \leq 1 \end{cases}
\end{aligned}
$$

Under this posterior distribution of the parameter θ, the probability of an accident occurring in the second period, $p(x_2 > \tau|x_1 \leq \tau)$, is given by

$$
p(x_2 > \tau|x_1 \leq \tau) = \int_\tau^1 \frac{\theta - \tau}{\theta} f(\theta|x_1 \leq \tau)d\theta = \frac{\tau - 1 - \log(\tau)}{1 - \log(\tau)}.
$$

As in Part 1, the probability derived from the posterior information is denoted as $p(\cdot|\boldsymbol{x})$. Therefore, at the start of the second period, the DM will decide to carry out the project only if

$$B - \frac{\tau - 1 - \log(\tau)}{1 - \log(\tau)} D^M \geq 0. \tag{9.1}$$

Next, examine the decision-making of the DM at the beginning of the first period. Under the no-information prior distribution, the probability of an accident occurring during the execution of the project in the first period, $p(x_1 > \tau|\boldsymbol{x}_0)$, is given by

$$p(x_1 > \tau|\boldsymbol{x}_0) = \int_{\tau}^{1} \frac{\theta - \tau}{\theta} d\theta = 1 - \tau + \tau \log(\tau). \tag{9.2}$$

Therefore, the value of the project is given by

$$\begin{cases} -D^M & \text{with probability } 1 - \tau(1 - \log(\tau)) \\ \max\left(B - \dfrac{\tau - 1 - \log(\tau)}{1 - \log(\tau)} D^M, 0\right) & \text{with probability } \tau(1 - \log(\tau)) \end{cases}$$

Hence, the expected project value is

$$\begin{aligned} &-(1 - \tau + \tau \log(\tau)) D^M \\ &\quad + \tau(1 - \log(\tau)) \max\left(B - \frac{\tau - 1 - \log(\tau)}{1 - \log(\tau)} D^M, 0\right). \end{aligned} \tag{9.3}$$

As a matter of course, for the value of Eq. (9.3) to be positive, the value of equation (9.1) must necessarily be positive. That is, if the decision to carry out the project is made at the beginning of the first period and it passes without an accident, the decision to stop the project at the beginning of the second period will not be made. In other words, the DM will either decide to stop the project in the first period considering the magnitude of the risk, carry out the project in the first period and encounter an accident, or execute the project in the first period without an accident and continue the project in the second period. If the project is carried out in the first period and passes without an accident, the fact of being accident-free leads to a revision of the expected value of the parameter of the random variable to a lower value. That is, the posterior expectation is revised in the direction of considering the project to be safer. Thus, the decision criterion for the second period will not be stricter than that of the first period. Therefore, there is no need for reconsideration in the second period.

Deciding to execute the project in period 1 essentially means making simultaneous decisions about the decision-making in period 2 and beyond. If the research and development period does not extend beyond period 1, the condition for the DM to promote the project, considering the probability of an accident occurring in the first phase (9.2), is as follows:

$$B - (1 - \tau + \tau \log(\tau))\, D^M > 0. \tag{9.4}$$

Meanwhile, when B/D^M is sufficiently large to satisfy (9.1), the condition for the expected project value (9.3) to be positive, assuming the project spans two periods, is

$$B\tau(1 - \log(\tau)) - D^M (1 - \tau)^2 = \tau\,(1 - \log(\tau)) \left(B - \frac{(1 - \tau)^2}{\tau\,(1 - \log(\tau))} D^M \right) > 0. \tag{9.5}$$

The condition (9.4) is more lenient relative to the condition for promoting the project in the first phase when there are two phases (9.5):

$$B - \frac{(1 - \tau)^2}{\tau(1 - \log(\tau))} D^M > 0.$$

The latter condition is naturally more stringent as it considers the risks of the second phase, based on an overall risk assessment of the entire route. Therefore, when deciding at the beginning of period 1 whether to promote the project, it is not rational to defer the decision on whether to promote the project itself and temporarily execute the project for only one phase, reassessing at the beginning of period 2. In this sense, deciding to execute the project in period 1 is equivalent to committing to decision-making in period 2 and beyond.

As described above, when it is not possible to observe the value of the random variable X, and thus learning is not possible, a decision must be made at the beginning of the period 1, evaluating the risk over the entire period covering periods 1 and 2. However, this poses an exceedingly difficult challenge for the DM or for society. Apart from having to make judgments using limited information, there is a tendency to assess long-term risks based on the immediate, short-term risks alone.

This is the problem pointed out by Samuelson (1963).[1] Samuelson considered a coin toss bet, with a probability of 0.5, where winning yields a profit of \$200 and losing incurs a loss of \$100. He discussed the rationality of the behavior of his colleague, who avoids this bet when it is offered only once, considering the \$100 loss a significant burden, but would accept the offer if it were for 100 bets. According to Samuelson, as long as one seeks to maximize expected utility, the rationality of a single bet should be equivalent to that of 100 bets.

The example discussed by Samuelson (1963) involves a coin toss bet and is based on the criterion of expected utility for judgment. Even so, in the decision-making for project execution, the criterion is expected profit. However, the structure of both

[1] See Rabin and Thaler (2001) for reference.

problems is the same. In the face of such risks, many insights have been obtained, especially through the development of behavioral economics and Prospect Theory regarding how humans make decisions in reality. In this context, the discussion is not originally based on maximizing expected utility, nor does it question the wisdom of decisions based on those discussions. Nevertheless, it is important to consider the notion of "framing," which suggests that even in cases of equivalent risk assessment, judgments can be significantly influenced by the differences in the set scenarios. In such cases, it is preferable to assess risks in accordance with long-term scenarios as much as possible, as judgments in short-term settings are easily influenced. However, this is not easy. If a decision is made to execute the project temporarily without assessing the overall risk at the beginning of period 1, it becomes equivalent to deciding to promote the project without properly assessing the whole risk. Regarding nuclear power generation, long-term risk assessment over the research and development process is desirable, but as Martin pointed out, projects tend to proceed based on short-term risk assessments that are relatively predictable. This leads to biased decision-making in favor of project execution.

References

Rabin, Matthew, and Richard H Thaler 2001."Anomalies: risk aversion." *Journal of Economic perspectives*. 15 (1): 219–232.

Samuelson, Paul 1963."Risk and uncertainty: A fallacy of large numbers." *scientia*. 57 (98): .

Chapter 10
Learning Through Trial and Error

This chapter examines scenarios where learning is possible, that is, knowledge will be accumulated throughout a project via trial and error. Even if an accident could potentially have been avoided, the process provides insight into the proximity of the risk, such as whether it could have been averted at the last moment or with a safety margin through the value of the precursor. This acquired information serves as valuable input for decision-making in subsequent periods. The chapter delves into the complexities of decision-making within such contexts. In the preceding chapter, we explored the strategy of initiating a project temporarily, monitoring its progress, and subsequently assessing the level of risk when learning is not possible. The current chapter examines the problems with this strategy in contexts where learning is possible.

In this chapter, instead of Assumption 9.10, we introduce

Assumption 10.1 All members of society can observe the value of the random variable X.

Consider the decision-making of the DM under Assumption 10.1. First, consider the situation at the beginning of period 2. The option to execute the project in period 2 is not available if it was not executed in period 1, which remains the same. Moreover, if the project was executed in period 1 and an accident occurred, it is still not possible to execute the project in period 2. Consider the case where the project was executed in period 1 and no accident occurred ($x_1 \leq \tau$). Under Assumption 9.9 of a no-information prior distribution, the probability density function of the posterior distribution of the parameter θ, given the observed value x_1 in the first phase, is denoted as $f(\theta|x_1)$. According to Bayes' theorem, we have

$$f(\theta|x_1) = \begin{cases} \dfrac{f(\theta)/\theta}{\displaystyle\int_{x_1}^{1} f(\theta')/\theta' d\theta'} = -\dfrac{1}{\theta \log(x_1)} & \text{for } x_1 \leq \theta \leq 1, \\ 0 & \text{otherwise.} \end{cases} \tag{10.1}$$

A. Torii, *Decision-making in High-Risk Projects*,
https://doi.org/10.1007/978-981-97-3056-8_10

Under this posterior distribution, the probability of an accident occurring in period 2 is

$$p(x_2 > \tau|x_1) = \int_{\tau}^{1} \frac{\theta - \tau}{\theta} f(\theta|x_1)d\theta = \frac{1 + \log(\tau) - \tau}{\log(x_1)}.$$

Therefore, the DM, at the beginning of period 2, will decide to execute the project only if

$$B - \frac{1 + \log(\tau) - \tau}{\log(x_1)} D^M > 0\,;$$

that is, only if

$$x_1 < \exp\left(\frac{D^M}{B}(1 + \log(\tau) - \tau)\right).$$

Denote the right-hand side of this inequality as $\tilde{x}_1$.

Note that the value of $\tilde{x}_1$ is not necessarily less than τ. As $\exp(1 - \tau + \log(\tau)) > \tau$ for $\tau \in [0, 1]$, the value of $\tilde{x}_1$ can exceed τ unless the ratio D^M/B is sufficiently large. Thus, even if an observed value of x_1 is close to τ, it does not necessarily mean the project should be stopped, as long as the risk is relatively low and the value of the project can be realized.

Therefore, in the following analysis, which focuses on how to decide to stop the project in situations where it should be stopped, only the case where $\tilde{x}_1 < \tau$ is considered. That is, we assume

Assumption 10.2

$$\frac{D^M}{B} > \frac{\log(\tau)}{1 - \tau + \log(\tau)}.$$

Consequently, the value of the project at the beginning of period 2 is given by

$$\max\left(B - \frac{1 + \log(\tau) - \tau}{\log(x_1)} D^M, 0\right) = \max\left(B\left(1 - \frac{\log(\tilde{x}_1)}{\log(x_1)}\right), 0\right). \tag{10.2}$$

Next, consider the DM's decision-making at the beginning of period 1. Under the no-information prior distribution, the probability density for the value x_1 taken by the random variable X is given by

$$\int_{x_1}^{1} \frac{1}{\theta} d\theta = -\log(x_1).$$

The value of the project is $-D^M$ when $x_1 \in (\tau, 1]$, and the value is given by (10.2) when $x_1 \in [0, \tau]$. Therefore, the expected value is

$$\begin{aligned}
&\int_0^{\tau} \max\left(B\left(1-\frac{\log(\tilde{x}_1)}{\log(x_1)}\right),0\right)(-\log(x_1))dx_1 \\
&\quad + \int_{\tau}^{1}(-D^M)(-\log(x_1))dx_1 \\
&\quad = \int_0^{\tilde{x}_1} B\left(1-\frac{\log(\tilde{x}_1)}{\log(x_1)}\right)(-\log(x_1))dx_1 + D^M\int_{\tau}^{1}\log(x_1)dx_1 \\
&\quad = B\tilde{x}_1 - D^M\left(1-\tau+\tau\log(\tau)\right).
\end{aligned}$$

Naturally, as shown below, under condition (9.1), this value is confirmed to be higher than the project value (9.5) in the case where the random variable X cannot be observed and no learning takes place.

$$\begin{aligned}
&\left(B\tilde{x}_1 - (1-\tau+\tau\log(\tau))D^M\right) - \left(B\tau(1-\log(\tau)) - D^M(1-\tau)^2\right) \\
&= B\left(\tilde{x}_1 - \tau + \tau\left(\log(\tau)-\log(\tilde{x}_1)\right)\right)
\end{aligned}$$

By considering the value of this equation as a function of $\tilde{x}_1$ and denoting it as $h(\tilde{x}_1)$, $h(\tau)=0$ and $h'(\tilde{x}_1) = B\left(1-\tau/\tilde{x}_1\right)$, leading to the conclusion that $h(\tilde{x}_1) \geq 0$ for $\tilde{x}_1 \leq \tau$.

Next, we investigate how the probability of executing or stopping the project changes depending on the actual value of the parameter θ. Given the value of θ, the random variable X follows a uniform distribution $U(0,\theta)$. The value of the project varies with θ as follows:

- When $\theta \in [0,\tilde{x}_1]$, the project is executed in period 2 at all realizations of the random variable X. No accident occurs and only the benefit B arises.
- When $\theta \in (\tilde{x}_1,\tau]$, the project is executed in period 2 and the benefit B arises only if the random variable X in the first period takes a value such that $x_1 \in [0,\tilde{x}_1]$. The probability of this occurring is $\tilde{x}_1/\theta$. No accident occurs either.
- When $\theta \in (\tau, 1]$, the project is executed in period 2 if the random variable X in period 1 takes a value such that $x_1 \in [0,\tilde{x}_1]$, with the probability of this occurring being $\tilde{x}_1/\theta$. Additionally, an accident occurs if the value of the random variable X in period 1 is $x_1 \in (\tau,1]$, with a probability of $1-\tau/\theta$, or if the value of X in period 1 is $x_1 \in [0,\tilde{x}_1]$, leading to the execution of the project in period 2, and the value of X in period 2 is $x_1 \in (\tau,1]$, again with a probability of $1-\tau/\theta$.

In summary, the project value depending on the value of θ is given by

$$\begin{cases}
B & \text{if } \theta \leq \tilde{x}_1, \\
\dfrac{\tilde{x}_1}{\theta}B & \text{if } \tilde{x}_1 < \theta \leq \tau, \\
\dfrac{\tilde{x}_1}{\theta}B - \left(1-\dfrac{\tau}{\theta}\right)\left(1+\dfrac{\tilde{x}_1}{\theta}\right)D^M & \text{if } \tau < \theta.
\end{cases}$$

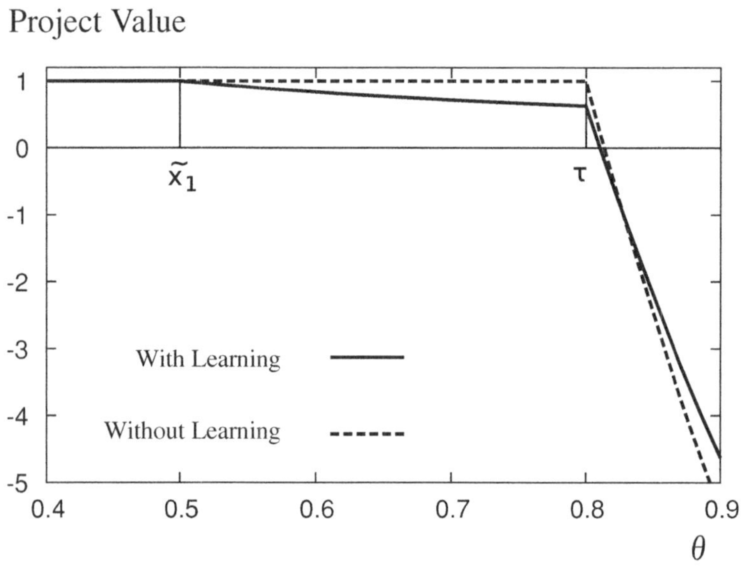

Fig. 10.1 Parameter θ and project value note: the case $\tau = 0.8$, $D^M/B = 30$, $\tilde{x}_1 = 0.499$

This is to be compared with the case where the value of the random variable X cannot be observed, and learning is not possible. We consider the scenario that the project is executed when the value of X is not observable. When $\theta \in [0, \tau]$, there is no chance of an accident; the project value is B. However, when $\theta \in (\tau, 1]$, the benefit is realized with the probability of avoiding an accident in the first period, which is τ/θ. Moreover, in this case, the probability that an accident occurs in either the first or second period is $1 - (\tau/\theta)^2$. Therefore, the project value is

$$\begin{cases} B & \text{if } \theta \leq \tau, \\ \dfrac{\tau}{\theta}B - \left(1 - \left(\dfrac{\tau}{\theta}\right)^2\right) D^M & \text{if } \tau < \theta. \end{cases}$$

A comparison is made between cases with and without learning. See Fig. 10.1. The realization of benefits is higher in the case without learning. In such a case, benefits are realized if no accidents occur in period 1 with probability τ/θ. However, in the case with learning, the project continues into period 2 and benefits are realized only if the value of the random variable X in period 1 is $x_1 \in [0, \tilde{x}_1]$ with probability $\tilde{x}_1/\theta$. However, the occurrence of accidents can be partially suppressed in the case with learning. In fact, when comparing the probabilities of accidents occurring in the two cases where $\theta > \tau$ (i.e., $1 - (\tau/\theta)^2$ for the case without learning and $(1 - \tau/\theta)(1 + \tilde{x}_1/\theta)$ for the case with learning), we find

$$1 - \left(\frac{\tau}{\theta}\right)^2 = \left(1 - \frac{\tau}{\theta}\right)\left(1 + \frac{\tau}{\theta}\right) > \left(1 - \frac{\tau}{\theta}\right)\left(1 + \frac{\tilde{x}_1}{\theta}\right).$$

In the case with learning, if the value of the random variable in period 1 is $x_1 \in (\tilde{x}_1, \tau]$, the project is halted, and the occurrence of accidents can be preemptively suppressed. This is reflected in the difference in the probabilities of accidents occurring.

As described above, the project value increases with learning because the project can be halted, and accidents can be prevented preemptively when the observed value of the random variable X falls within a specific range. The significance of groping-in-the-dark lies in possessing an algorithm to stop the project in this manner.

However, appropriately "halting" high-risk projects is challenging. In cases where learning is not possible, there is a challenge in assessing the long-term risk across the entire trajectory. Even in cases where learning is possible, it is important to note that not all accidents can be prevented. In situations where there is a possibility of accidents occurring, with $\theta \in (\tau, 1]$, the project is halted only in some cases. Even dangerous projects will continue to be executed if $x_1 \leq \tilde{x}_1$.

Furthermore, upon closer examination of the halting mechanism discussed here, it becomes apparent that higher risk projects are, in fact, less likely to be halted. Dangerous projects (i.e., those with $\theta \in (\tau, 1]$,) are halted based on the observed value of the random variable X in period 1, specifically when $x_1 \in (\tilde{x}_1, \tau]$. The probability of this occurring is $(\tau - \tilde{x}_1)/\theta$, which is a decreasing function of θ. The probability of halting decreases as the risk increases because, although the probability of the project proceeding, represented by $\tilde{x}_1/\theta$, becomes lower, the likelihood of an accident occurring in the first period increases. Thus, the probability of appropriately halting the project decreases. (See Fig. 10.2 for details.)

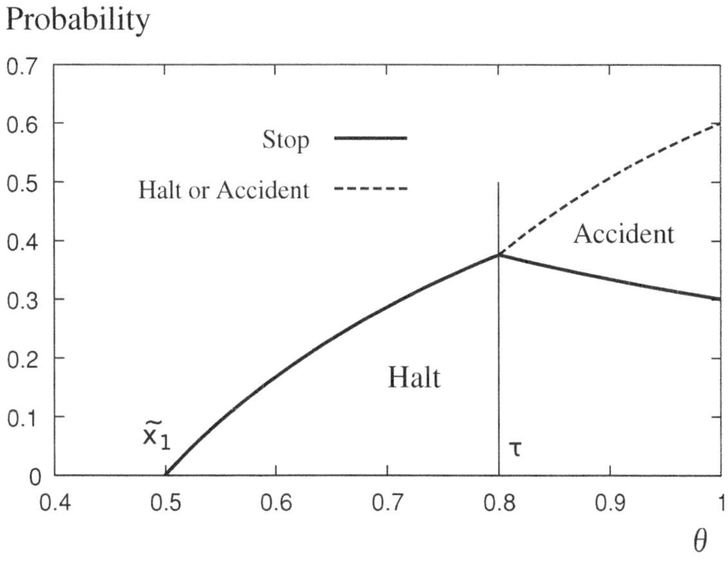

Fig. 10.2 The Relationship between the parameter θ and the probabilities of project halting and accident occurrence note: the case $\tau = 0.8$, $D^M/B = 30$, $\tilde{x}_1 = 0.499$

Chapter 11
Conclusion of Part 2

This part explored what HROs mean in the context of research and development by creating and analyzing a model suited for hazardous R&D activities. It focused on incorporating information indicative of project risks during the R&D process, interpreting this information optimally, and considering organizational responses, including project termination, to mitigate risks.

In the model without learning, discussed in Sect. 9.2, the decision to proceed with the project in the first period is equivalent to committing to the decision in the second period. In this sense, a system without learning, as analyzed here, is tightly coupled. Conversely, in the model with decision-making involving learning, analyzed in Chapter 11, the decision in the first period does not dictate the decision in the second period. At the beginning of the second period, a decision is made anew using the information obtained in the first period. Hence, the coupling is loose. Meanwhile, since the risk is successfully reduced through learning, it can be considered an example of a HRO. Perrow (1999) suggests that HROs are not necessarily about managing risk in tightly coupled systems but about creating systems with loose coupling. The system with learning presented here aligns with this interpretation.

Notably, the model in this paper is designed with the R&D process in mind. Reducing risk in projects involving R&D is achieved by incorporating a mechanism to halt the system when signs of danger appear. Moreover, halting the project's execution is interpreted as creating loose coupling. However, the discussion of general HROs typically focuses on learning in regular operations, where the ability to halt is not necessarily the source of risk reduction.

Furthermore, this paper assumes that halting R&D operations midway can avert risk. However, it may also increase risk. As noted by Martin (2012), after the Three Mile Island nuclear power plant accident, the stagnation in R&D activities related to nuclear power led to the widespread adoption of riskier second-generation nuclear reactors. In the case of nuclear power, even if a project is halted, issues remain, such as decommissioning and handling spent nuclear fuel and high-level radioactive waste disposal. For instance, it is uncertain how to best halt the nuclear power system at this point to reduce risk. Perhaps we are like a dog or cat, hesitating while trapped in the middle of a busy road.

A. Torii, *Decision-making in High-Risk Projects*,
https://doi.org/10.1007/978-981-97-3056-8_11

References

Martin, Richard 2012. *Superfuel: thorium, the green energy source for the future.* : St. Martin's Press.

Perrow, Charles 1999. *Normal accidents: Living with high risk technologies.* : Princeton university press.

SPRINGER NATURE

GPSR Compliance

The European Union's (EU) General Product Safety Regulation (GPSR) is a set of rules that requires consumer products to be safe and our obligations to ensure this.

If you have any concerns about our products, you can contact us on ProductSafety@springernature.com

In case Publisher is established outside the EU, the EU authorized representative is:

Springer Nature Customer Service Center GmbH
Europaplatz 3
69115 Heidelberg, Germany

The manufacturer's authorised representative in the EU is Springer Nature Customer Service Centre GmbH, Europaplatz 3, 69115 Heidelberg, Germany. If you have any concerns regarding our products, please contact ProductSafety@springernature.com

Printed and bound by CPI Group (UK) Ltd, Croydon, CR0 4YY
01/10/2026
02225876-0004